Contents

Dedications and acknwledgements

I should like to dedicate this book to my late father, grandmothers, grandfathers and uncles Petr and Ivan.

I should also like to express my special thanks to my mother.

Introduction

The conception of *equivalential* universal Horn theory goes back — within the context of *sentential* languages (viz., first-order languages with the single unary relation symbol D; cf. [4],[14]) — to [19]. It has been specified and studied — within the general context of universal Horn theories over arbitrary first-order languages — in [34]. Though the latter work provides a quite advanced study of such UHT (especially, upon the basis of the general theory of equivalence between UHT developed therein), there are still several issues, which have proved beyond the mentioned study, while they appear quite acute in exploring many particular UHTs. This reason has motivated writing this additional book that incorporates not merely advancing certain general issues of [34] but also extensive exemplifying this advance by considering several intersting particular UHTs appearing in the literature on Logic.

The key paradigm of the current study is that equivalential UHTs can be viewed as algebraizable but with equivalent algebraic semantics (viz., algebraic counterpart) constituted by rather algebraic systems (more precisely, all simple models of the equivalential UHT under consideration) than pure algebras (more precisely, the underlying algebras of all simple models of the algebraizable UHT under consideration). This is actually a particular case of Mal'cev's generic paradigm "algebraic systems versus pure algebras" (cf. [16]). The former can then be formulated as "equivalential UHTs versus algebraizable ones". The current book is entirely devoted to advancing this thesis.

The rest of the monograph is as follows. Chapter 1 is an advance of certain general issues elaborated in [34]. Section 1.1 is a concise self-contained summary of general set-theoretical and algebraic background needed for comprehending the present book in its own right. Further, in Section 1.2 we extend to infinitary Horn Logic and arbitrary (viz., not necessarily purely translations) those results of [24] which deal with extensions of equivalent UHTs.[1] The

[1] These issues are especially acute for finding deductive bases (viz., axiomatizations) as well as both complete and sound semantics of calcili constituted by all rules admissible in given

thing is that, though the general study [24] with its generic first-order formalism is well-applicable to *finitely*-equivalential UHT, as it has been shown in Section 4.2 of [34], there are quite interesting particular equivalential (and what is more, algebraizable) finitary finite UHTs over a sentential language, which are not *finitely*-equivalential (and so, are not *finitely*-algebraizable). For this reason, the solely finitary context of [24] would be far from being sufficient for our generic purpose. Next, Section 1.3 is a concise summary of the conception of rational equivalence between prevarieties that has actually been extended from pure algebras [15] to algebraic systems in [34]. We specify this issue here explicitly. Section 1.4 equally makes explicit the correspondence between subprevarieties with their relative axiomatizations and extensions of appropriate equational theories together with their relative axiomatizations essentially discovered in Section 1.4 of [34]. In Section 1.5 we extend the conception of equivalent algebraic semantics defined in [34] (to be called here *purely*-algebraic) from pure algebras to algebraic systems. Combining main results of Sections 1.2 and 1.4, we obtain extensions of those results of [24], which concern correspondence between extensions of algebraizable finitary UHTs and subquasivarieties of their equivalent purely-algebraic semantics, to both infinitary Logic, equivalential UHTs, algebraic systems and arbitrary translations. Section 1.6 is, in part, a brief summary of those results of [34] which, in particular, argue that the class of all simple models of an equivalential UHT is a unique (up to rational equivalence) equivalent algebraic semantics for it. More generally, Section 1.6 justifies our main thesis "equivalential versus algebraizable UHTs". More precisely, we prove that an UHT is equivalential iff it has an equivalent algebraic semantics, while, by definition, an UHT is algebraizable iff it has an equivalent purely-algebraic semantics.

Chapter 2 incorporates a summary of necessary issues of the approach of the work [31] that has proposed a generic constructive method of association of multiple-conclusion sequent calculi with finitely-valued logics with *equality determinant* and advanced towards General Algebraic Logic in [32] (cf. [33]), where it has been proved that all the sequent calculi with structural rules suggested in [31] are finitely-equivalential, in which case they are covered by the current work. We argue that, under presence of lattice conjunction and disjunction connectives, simple models of such calculi are rationally equivalent (in our extended sense) to *quasi-ordered algebras* that are algebraic systems with underlying algebra, being a lattice, and with a single binary relation being a quasi-ordering properly compatible with lattice conjunction and disjunction

sequent calculi.

like partial orderings of lattices are compatible with lattice operations.

In Chapter 3 we explore several particular finitely-valued logics with equality determinant, conjunction and disjunction with non-algebraizable associated sequent calculi. It includes algebraically-advanced studies of related quasi-ordered algebras, collectively showing that equivalent algebraic semantics of non-algebraizable equivalential UHTs are their algebraic counterparts of full value, to which advanced tools of Universal Algebra are well-applicable. This well justifies our key thesis "equivalential versus algebraizable UHTs" underlying the current book and being a particular case of Mal'cev's generic paradigm "algebraic systems versus pure algebras" [16]. In this connection, it is remarkable that Section 3.3 is devoted to a sentential logic equivalent to the associated sequent calculus. In this way, applicability of Chapter 1 goes beyond the context of Chapter 2.

Conclusions provide a concise purely-descriptive summary of principal contributions of the book.

Appendix A is an appropriate extensions of certain results of Section 2.1 of [34] to arbitrary translations, without which main results of Chapter 1 could not be obtained within the general context of arbitrary translations. Section A.1 advances the issue of Deduction Theorem (DT) for universal Horn theories that is due to Appendix B of [35]. More precisely, we extend the issue of preservation of DT under equivalence between finitary UHT (cf. Theorem B.2 of [35]) to arbitrary translations (cf. Theorem A.6).

Appendices B and C provide supplementary generic tools of proper constructing fragmentary versions (in particular, single-premise and/or single-conclusion ones) of associated sequent calculi.

The composition of the material is chosen to be self-contained. Unless otherwise specified, we entirely follow the customary conventions adopted in [24], [34] and [35]. For general basic issues concerning Lattice Theory, Universal Algebra and Model Theory, the reader is also referred to standard mathematical handbooks like [2], [5],[6], [9], [10], [11] and, especially, [16].

The present monograph incorporates those results of mine, which were obtained during my research starting from my post-graduated study since 1991 and which have proved beyond the scopes of my previous books [34] and [35].

On the other hand, the present book together with both [34] and [35] constitute foundations of General Algebraic Logic.

And what is more, this monograph starts a "new wave" in Algebraic Logic consisting in association of rather algebraic systems than pure algebras with logical systems. This expands scopes of Algebraic Logic much, once the conception of being equivalential is much wider than that of being algebraizable.

And concrete examples, explored in Chapter 3, constitute just a little portion of representatives of such expansion.

In this connection, it is remarkable that equivalence between UHT preserves not merely extensions but equally many other metalogical issues. For example, it preserves Deduction Theorem (cf. Appendix B of [35]). In this way, applicability of novel scopes of Algebraic Logic, expanded by this work, goes far beyond the context of the present monograph.

Thus, the general first-order syntactical formalism of [24] (as opposed to the too-restricted "many-dimensional" one adopted, e.g., in [26]), initially involved just to cover sequent calculi of any kind (see also [33] in this connection), though being definitely valuable in its own right for such a specific purpose, has eventually proved crucial for providing the revolutionary innovation described above.

This demonstrates quite unexpected ways of development of fundamental science.

Chapter 1

General issues

1.1 Preliminaries

We follow the conventions adopted in [34]. Unless otherwise specified, we fix an arbitrary first-order signature $L = \langle F, R \rangle$. The arity of any symbol $s \in F \cup R$ in L is denoted by $\mu_L(s)$. For unifying notations, L-structures are denoted by Calligraphic letters $\mathcal{A}, \mathcal{B}, \mathcal{C}, \ldots$ (possibly, with indices), while [their underlying] F-algebras are denoted by [corresponding] Fraktur letters $\mathfrak{A}, \mathfrak{B}, \mathfrak{C}, \ldots$ (with [the same] indices if any), whereas their carriers are denoted by respective capital Italic letters $A, B, C, \ldots$ (with respective indices if any). The class of all L-structures is denoted by S_L.

Natural numbers (including 0) are treated as ordinals, i.e., sets constituted by lesser natural numbers, the countable ordinal of all them being denoted by ω. The class of all ordinals is denoted by ∞. (The expression $\alpha < \infty$ means $\alpha \in \infty$.) During the book, κ is supposed to be an arbitrary regular cardinal (in particular, ω).

Given any $\overline{s} \in (F \cup R)^n$, where $n \in \omega$, put $\mu_L(\overline{s}) \triangleq \sum_{i \in n} \mu_L(s_i)$. Script letters x, y, z (possibly, with indices) are used for denoting variables. Given any $\alpha \leqslant \infty$, put $V^\alpha \triangleq \{x_\beta\}_{\beta \in \alpha}$. The class of all F-terms (equality-free atomic L-formulas) with variables in V^α is denoted by $\mathrm{Tm}_F(V^\alpha)$ (resp., by $\mathrm{At}_L(V^\alpha)$). Given any $n \in \omega$ and $\alpha \in \infty$, set $\overline{x}(\alpha, n) \triangleq (x_{\alpha+i})_{i \in n} \in (V^\infty)^n$ and $\overline{x}(n) \triangleq \overline{x}(0, n)$. The set of all variables really occurring in any universal sentence[1] S is denoted by $\mathrm{Var}(S)$.

Let A be a set. A *quasi-ordering on A* is any reflexive and transitive binary relation on A. An *equivalence on A* is any symmetric quasi-ordering E on A, in

[1]As usual, universal prefixes are normally omitted.

which case we have the *natural mapping* $\nu_E : A \to A/E \triangleq \{[a]_E | a \in A\}, a \mapsto [a]_E \triangleq \{b \in A | aEb\}$. An example of an equivalence on A is the *diagonal relation* (viz., *identity mapping*) $\Delta_A \triangleq \{\langle a, a\rangle | a \in A\}$ *on* A. In addition, given any mapping h from A to any set B, its *kernel* $\ker h \triangleq \{\langle a, b\rangle \in A^2 | ha = hb\}$ is an equivalence on A. Moreover, $\ker \nu_E = E$.

Let $X \subseteq A$. Then, we have both the quasi-ordering $\varpi_A^X \triangleq \{\langle a, b\rangle \in A^2 | a \in X \Rightarrow b \in X\}$ on A and the *characteristic function* χ_A^X *of* X *in* A being the mapping from A to 2 such that $\chi_A^X[X] = \{1\}$ and $\chi_A^X[A \setminus X] = \{0\}$.

Given three sets A, B, C and two mappings $f : A \to B$ and $h : A \to C$, we have $(f \times h) : A \to (B \times C), a \mapsto \langle fa, ha\rangle$.

Let $\mathfrak{A}$ be an F-algebra. A *congruence of* $\mathfrak{A}$ is any equivalence θ on A such that, for each $f \in F$ and all $\bar{a}, \bar{b} \in A^{\mu_L(f)}$ such that $a_i \theta b_i$, for every $i \in \mu_L(f)$, it holds that $f^{\mathfrak{A}}(\bar{a}) \theta f^{\mathfrak{A}}(\bar{b})$, the set of all them being denoted by $\mathrm{Con}(\mathfrak{A})$. Given any $\theta \in \mathrm{Con}(\mathfrak{A})$, we have the *quotient* $\mathfrak{A}/\theta \triangleq \langle A/\theta, \nu_\theta \circ f^{\mathfrak{A}} \circ (\nu_\theta)^{-1}\rangle_{f \in F}$ *of* $\mathfrak{A}$ *by* θ. Given a one more F-algebra $\mathfrak{B}$, a *homomorphism from* $\mathfrak{A}$ *to* $\mathfrak{B}$ is any $h : A \to B$ such that, for each $f \in F$ and every $\bar{a} \in A^{\mu_L(f)}$, it holds that $hf^{\mathfrak{A}}(\bar{a}) = f^{\mathfrak{B}}(h\bar{a})$, in which case $\ker h \in \mathrm{Con}(\mathfrak{A})$, the set of all them being denoted by $\mathrm{Hom}(\mathfrak{A}, \mathfrak{B})$. Note that, for any $\theta \in \mathrm{Con}(\mathfrak{A})$, $\nu_\theta \in \mathrm{Hom}(\mathfrak{A}, \mathfrak{A}/\theta)$.

Let $\mathcal{A}$ be an L-structure. A *congruence of* $\mathcal{A}$ is any $\theta \in \mathrm{Con}(\mathfrak{A})$ such that, for each $r \in R$ and all $\bar{a}, \bar{b} \in A^{\mu_L(r)}$ such that $a_i \theta b_i$, for every $i \in \mu_L(r)$, it holds that $\bar{a} \in r^{\mathfrak{A}} \Leftrightarrow \bar{b} \in r^{\mathfrak{A}}$, the set of all them being denoted by $\mathrm{Con}(\mathcal{A})$. According to Proposition 1.4 of [34], one can equivalently set $\Im(\mathcal{A}) \triangleq \bigcup \mathrm{Con}(\mathcal{A}) \in \mathrm{Con}(\mathcal{A})$. Then, $\mathcal{A}$ is said to be *simple*, if $\Im(\mathcal{A}) = \Delta_A$. Given any $\theta \in \mathrm{Con}(\mathfrak{A})$, we have the *quotient* $\mathcal{A}/\theta \triangleq \langle \mathfrak{A}/\theta, \nu_\theta[r^{\mathcal{A}}]\rangle_{r \in R}$ *of* $\mathcal{A}$ *by* θ. Given a one more L-structure $\mathcal{B}$, a *(strict) homomorphism from* $\mathcal{A}$ *to* $\mathcal{B}$ is any $h \in \mathrm{Hom}(\mathfrak{A}, \mathfrak{B})$ such that, for each $r \in R$, $r^{\mathcal{A}} \subseteq (=) h^{-1}[r^{\mathcal{B}}]$ (in which case $\ker h \in \mathrm{Con}(\mathcal{A})$), the set of all them being denoted by $\mathrm{Hom}(\mathcal{A}, \mathcal{B})$ (resp., by $\mathrm{SHom}(\mathcal{A}, \mathcal{B})$). Note that, for any $\theta \in \mathrm{Con}(\mathfrak{A}) (\in \mathrm{Con}(\mathcal{A}))$, $\nu_\theta \in \mathrm{Hom}(\mathcal{A}, \mathcal{A}/\theta)$ (resp, $\nu_\theta \in \mathrm{SHom}(\mathcal{A}, \mathcal{A}/\theta)$). Injective (and surjective) strict homomorphims are referred to as *embeddings* (resp., *isomorphisms*). Given an F-algebra $\mathfrak{A}$, an L-structure $\mathcal{B}$ and any $h \in \mathrm{Hom}(\mathfrak{A}, \mathfrak{B})$, we have the L-structure $h^{-1}[\mathcal{B}] \triangleq \langle \mathfrak{A}, h^{-1}[r^{\mathcal{B}}]\rangle_{r \in R}$. In case $h = \Delta_A$, we set $\mathcal{B} \upharpoonright A \triangleq h^{-1}[\mathcal{B}]$. Structures of such a form are referred to as *substructures of* $\mathcal{B}$.

Put $(L + \approx) \triangleq \langle F, R + \approx\rangle$, where $(R + \approx) \triangleq R \cup \{\approx\}$, while $\mu_{L+\approx}$ is the extension of μ_L to $F \cup (R + \approx)$ defined by $\mu_{L+\approx}(\approx) \triangleq 2$. Any $(L + \approx)$-structure $\mathcal{A}$ is naturally identified with the couple $\langle \mathcal{A} \upharpoonright L, \approx^{\mathcal{A}}\rangle$. Then, given an L-structure $\mathcal{A}$, we have the $(L + \approx)$-structure $(\mathcal{A} + \approx) \triangleq \langle \mathcal{A}, \Delta_A\rangle$. Given a class of L-structures K, set $(\mathsf{K} + \approx) \triangleq \{\mathcal{A} + \approx | \mathcal{A} \in \mathsf{K}\}$. Recall that $\vdash_{\mathsf{K}}$ denotes the class of all *equality-free* infinitary universal Horn L-sentences (viz.

-*rules* or -*implications*) true in each member of K. In this way, $\vdash^{\approx}_{\mathsf{K}} \triangleq \vdash_{\mathsf{K}+\approx}$ does actually involve sentences with equality. Given an equality-free (κ-*ary*) universal Horn L-theory $\mathbb{T}$ (that is, such that every $(\Gamma \to \Phi) \in \mathbb{T}$ is κ-*ary*, i.e., $|\Gamma| < \kappa$).[2] by $\mathrm{Mod}(\mathbb{T})$ we denote the class of all *models of* $\mathbb{T}$, that is, those L-structures, which satisfy all sentences in $\mathbb{T}$, in which case, according to Theorem 1.6 of [34], one can equivalently set $\vdash_{\mathbb{T}} \triangleq \vdash_{\mathrm{Mod}(\mathbb{T})}$, rules in which are referred to as *derivable in* $\mathbb{T}$, according to Appendix A of [35]. By $\mathrm{Mod}_{\Im}(\mathbb{T})$ we denote the class of all simple models of $\mathbb{T}$. An *extension of* $\mathbb{T}$ is any UHT of the form $\vdash_{\mathbb{T}'} \supseteq \mathbb{T}$, in which case it is said to be *(relatively) axiomatized by* $\mathbb{T}'$ *(by* $\mathbb{T}' \setminus \mathbb{T}$, respectively). An extension is referred to as *axiomatic*, whenever it is relatively axiomatized by a class of L-*axioms* (viz., premise-less L-rules). An L-rule R is said to be *admissible in* $\mathbb{T}$, if $\mathbb{T}$ and $\mathbb{T} \cup \{R\}$ have the same derivable axioms. Then, $\mathbb{T}$ is said to be *structurally complete*, if any L-rule being admissible in $\mathbb{T}$ is derivable in it. The class $\mathrm{Adm}(\mathbb{T})$ of all L-rules admissible in $\mathbb{T}$ is structurally complete and is the greatest extension of $\mathbb{T}$ with the same derivable axioms. Next, given any (κ-ary) universal Horn $(L+\approx)$-theory $\mathbb{T}$, put $\mathrm{Mod}_{\approx}(\mathbb{T}) \triangleq \{\mathcal{A} \in \mathsf{S}_L | (\mathcal{A}+\approx) \in \mathrm{Mod}(\mathbb{T})\}$ (this is the class of all models of $\mathbb{T}$ in the standard interpretation of Model Theory with equality). Then, a *(κ-ary) prevariety* (viz., *implicational class*) is any class of such a kind, in which case it is said to be *axiomatized by* $\mathbb{T}$. (Finitary prevarieties are referred to as *quasi-varieties*, while finitary rules are also called *quasi-identities*.) A *variety* is any prevariety axiomatized by an UHT consisting of identities alone. The prevariety *generated by* $\mathsf{K} \subseteq \mathsf{S}_L$ is defined to be the one axiomatized by $\vdash^{\approx}_{\mathsf{K}}$. (It is the least one including K.)

An L-structure $\mathcal{A}$ is said to be *singular* (viz., *one-element*), whenever $|A| = 1$. In that case, is is said to be *trivial*, provided, in addition, for each $r \in R$, $r^{\mathcal{A}} = A^{\mu_L(r)}$. Thus, $\mathcal{A}$ is trivial iff, for each $r \in (R+\approx)$, $r^{\mathfrak{A}+\approx} = A^{\mu_{L+\approx}(r)}$.

Let K be a class of L-structures. By $\mathbf{NT}(\mathsf{K})$ (resp. by $\mathbf{S}(\mathsf{K})$)we denote the class of all non-trivial (substructures of) members of K. Further, by $\mathbf{Alg}(\mathsf{K})$ we denote the category constituted by members of K as objects and by homomorphisms between them as morphisms. The implicational class generated by K is denoted by $\mathbf{I}(\mathsf{K})$. According to [16], an L-structure $\mathcal{A} \in \mathbf{I}(\mathsf{K})$ iff $\mathcal{A}$ is isomorphic to a substructure of a direct product of members of K or, equivalently, $\bigcap\{\ker h | h \in \mathrm{Hom}(\mathcal{A},\mathcal{B}), \mathcal{B} \in \mathsf{K}\} = \Delta_A$ and $\bigcap\{h^{-1}[r^{\mathcal{B}}] | h \in \mathrm{Hom}(\mathcal{A},\mathcal{B}), \mathcal{B} \in \mathsf{K}\} = r^{\mathcal{A}}$, for each $r \in R$, or, equivalently, $\bigcap\{h^{-1}[r^{\mathcal{B}+\approx}] | h \in \mathrm{Hom}(\mathcal{A},\mathcal{B}), \mathcal{B} \in \mathsf{K}\} = r^{\mathcal{A}+\approx}$, for each $r \in R+\approx$. Clearly:

$$\mathbf{I}(\mathsf{K}) = \mathbf{I}(\mathbf{NT}(\mathsf{K})). \tag{1.1}$$

[2] As usual, we say "finitary" for "ω-ary" within any context.

A subclass of K is said to be *relatively axiomatized by* a universal Horn $(L+\approx)$-theory $\mathbb{R}$ if it is equal to $\mathsf{K} \cap \mathrm{Mod}_{\approx}(\mathbb{R})$. A *relative subvariety of* K is any subclass of it relatively axiomatized by a set of L-axioms (viz., L-identities) $\mathbb{A}$, in which case it is closed under formation of substructures, isomorphic copies, direct products as well as quotients belonging to K. In particular, by (1.1), we have:

$$\mathbf{I}(\mathsf{K}) \cap \mathrm{Mod}_{\approx}(\mathbb{A}) = \mathbf{I}(\mathbf{S}(\mathsf{K}) \cap \mathrm{Mod}_{\approx}(\mathbb{A})). \tag{1.2}$$

The *variety generated by* K, that is, the variety axiomatized by all identities true in K is denoted by $\mathbf{V}(\mathsf{K})$. A variety V is said to be K-*open* (cf. [21], [24]), provided $\mathsf{V} = \mathbf{V}(\mathsf{K} \cap \mathsf{V})$.

Remark 1.1. Clearly, the mappings $\mathsf{R} \mapsto \mathbf{V}(\mathsf{R})$ and $\mathsf{V} \mapsto (\mathsf{K} \cap \mathsf{V})$ are inverse to one another isomorphisms between the complete lattices of all relative subvarieties of K and all K-open subvarieties of $\mathbf{V}(\mathsf{K})$. ∎

A prevariety P is said to be *structurally complete* (cf. [24]), if, for every subprevariety $\mathsf{P}' \subseteq \mathsf{P}$, $\mathbf{V}(\mathsf{P}') = \mathbf{V}(\mathsf{P}) \Rightarrow \mathsf{P}' = \mathsf{P}$.

Proposition 1.2. *Let $\mathcal{A}$ be an L-structure. Then, the following are equivalent:*

(i) *each singular substructure of $\mathcal{A}$ is trivial;*

(ii) *every singular member of $\mathbf{I}(\mathcal{A})$ is trivial.*

Proof. (ii)⇒(i) is trivial, for substructures of $\mathcal{A}$ belong to $\mathbf{I}(\mathcal{A})$. Conversely, assume (i) holds. Take any singular $\mathcal{B} \in \mathbf{I}(\mathcal{A})$. Consider any $h \in \mathrm{Hom}(\mathcal{B}, \mathcal{A})$. Then, $\mathcal{A} \upharpoonright h[B]$ is a singular substructure of $\mathcal{A}$, so, by (i), it is trivial, in which case $h^{-1}[\mathcal{A}]$ is trivial as well. Thus, $\mathcal{B}$ is trivial. In this way, (ii) holds, as required. ∎

1.2 Extensions of equivalent universal Horn theories

During the rest of this chapter, we follow the notation conventions adopted in Chapter 2 of [34]. As for the notion of translation and related issues (that are too cumbersome to reproduce them elsewhere), the reader is referred directly to Chapter 2 of [34].

Theorem 1.3. *Suppose $\mathbb{T}_1$ and $\mathbb{T}_2$ are equivalent with respect to (L_1, L_2)-translation $\tau_{1,2}$ and (L_2, L_1)-translation $\tau_{2,1}$. Then, the following hold:*

(i) $(\tau_{2,1})^{-1}$ *and* $(\tau_{1,2})^{-1}$ *are inverse to one another isomorphims between the lattices of all extensions of* $\mathbb{T}_1$ *and* $\mathbb{T}_2$*, corresponding extensions being equivalent with respect to translations involved;*

(ii) *for each* $i \in 3 \setminus 1$ *and every universal Horn* L_i*-theory* $\mathbb{R}$*:*[3]

$$\vdash_{\mathbb{T}_i \cup \mathbb{R}} = (\tau_{i,3-i})^{-1}[\vdash_{\mathbb{T}_{3-i} \cup \tau_{i,3-i}[\mathbb{R}]}];$$

(iii) $(\tau_{2,1})^{-1}$ *and* $(\tau_{1,2})^{-1}$ *are inverse to one another isomorphims between the lattices of all axiomatic extensions of* $\mathbb{T}_1$ *and* $\mathbb{T}_2$*, the corresponding axiomatic extensions being equivalent with respect to translations involved;*

(iv) $\mathbb{T}_1$ *is structurally complete iff* $\mathbb{T}_2$ *is.*

Proof. (i) Let $i \in 2$. Consider any extension $\mathbb{X}$ of $\mathbb{T}_i$. Then, the conditions 2.1(v) and 2.7(v) being valid for $\vdash_i$ (in view of Theorem 2.7 of [34]) remain true for $\mathbb{X}$, in which case, by Lemma A.2, $(\tau_{1-i,i})^{-1}[\mathbb{X}]$ is an extension of $\mathbb{T}_{1-i}$. The fact that the mappings under consideration are order-preserving is evident. Further, with using Definition 2.1(iii,iv) of [34], inherited by arbitrary extensions of $\mathbb{T}_1$ and $\mathbb{T}_2$, it is routine checking that the mappings involved are inverse to one another. Finally, the conditions 2.1(iii,vi) of [34] being valid for $\mathbb{T}_1$ and $\mathbb{T}_2$ remain true for their arbitrary extensions. After all, 2.1(i,ii) of [34] hold for corresponding extensions by definitions of the mappings $(\tau_{2,1})^{-1}$ and $(\tau_{1,2})^{-1}$. This completes the argument of (i).

Next, (ii) immediately follows from (i) and Corollary A.5. Further, (iii) immediately follows from (i) and the particular case of (ii), when $\mathbb{R}$ consists of axioms alone. Finally, (iv) is an immediate consequence of (i). ■

The particular finitary purely relational case of Theorem 1.3 is due to [24].

1.3 Rationally equivalent classes of algebraic systems

Given a class of similar algebraic systems K, we have the injection $\mho_{\mathsf{K}} : \mathsf{K} \to (\mathsf{K} + \approx), \mathcal{A} \mapsto (\mathcal{A} + \approx)$.

[3]This item means that the isomorphisms mentioned in the item (i) preserve relative axiomatizations.

Let K_i, where $i \in 3 \setminus 1$, be a class of L_i-structures.[4] Then, K_1 and K_2 are said to be *(L_0-conservatively [functionally* or *relationally]) rationally equivalent* {*with respect to* (L_0-conservative [resp., purely functional or relational]) $(L_1, L_2 + \approx)$-translation τ *and* $(L_2, L_1 + \approx)$-translation ρ}, provided τ is compatible with $\mathsf{K}_2 + \approx$, ρ is compatible with $\mathsf{K}_1 + \approx$, while the mappings $\tau^{-1} \circ \mho_2$ and $\rho^{-1} \circ \mho_1$ are inverse to one another bijections between K_2 and K_1. This extends Mal'cev's conception of rational equivalence [15] defined therein for algebras and purely functional translations.

Given an $(L_1, L_2 + \approx)$-translation τ, by $\tau^{\approx}$ we denote the $(L_1 + \approx, L_2 + \approx)$-translation being the extension of τ to $L_1 + \approx$ defined by $\tau(\approx) \triangleq \{x_0 \approx x_1\}$. Then, given any L_2-structure $\mathcal{A}$ compatible with τ, $\mathcal{A} + \approx$ is compatible with both τ and $\tau^{\approx}$, while:

$$(\tau^{\approx})^{-1}[\mathcal{A} + \approx] = \tau^{-1}[\mathcal{A} + \approx] + \approx. \tag{1.3}$$

Thus, Theorem 2.10(v)⇒(i), Corollary 2.12, (1.14), (1.15) and (2.12) of [34] collectively with (1.3) immediately yield:

Theorem 1.4. *Let* K_1 *and* K_2 *be as above. Then,* $\vdash_1^{\approx}$ *and* $\vdash_2^{\approx}$ *are* L_0*-conservatively equivalent iff they are* $(L_0 + \approx)$*-conservatively equivalent. More precisely, it holds that (n)⇒(ii)⇔(iii), where n is either "i" or "iv", while:*

(i) $\vdash_1^{\approx}$ *and* $\vdash_2^{\approx}$ *are equivalent with respect to an* $(L_1 + \approx, L_2 + \approx)$*-translation* τ *and an* $(L_2 + \approx, L_1 + \approx)$*-translation* ρ*;*

(ii) $\vdash_1^{\approx}$ *and* $\vdash_2^{\approx}$ *are equivalent with respect to the* $(L_1 + \approx, L_2 + \approx)$*-translation* $(\tau \upharpoonright L_1)^{\approx}$ *and the* $(L_2 + \approx, L_1 + \approx)$*-translation* $(\rho \upharpoonright L_2)^{\approx}$*;*

(iii) $\mathbf{I}(\mathsf{K}_1)$ *and* $\mathbf{I}(\mathsf{K}_2)$ *are rationally equivalent with respect to* $(\tau \upharpoonright L_1)$ *and* $(\rho \upharpoonright L_2)$*;*

(iv) K_1 *and* K_2 *are rationally equivalent with respect to* $(\tau \upharpoonright L_1)$ *and* $(\rho \upharpoonright L_2)$*;*

Theorem 1.4(i)⇒(ii)⇔(iii) extends Corollary 2.13 of [34] to algebraic systems. On the other hand, according to (1.14), (1.15) and Corollary 2.23 of [34], we, in addition, get:

Corollary 1.5. *Let* P_i*, where* $i \in 3 \setminus 1$*, is a prevariety of* L_i*-structures. Then,* P_1 *and* P_2 *are* L_0*-conservatively rationally equivalent iff there is an isofunctor between* $\mathbf{Alg}(\mathsf{P}_1)$ *and* $\mathbf{Alg}(\mathsf{P}_2)$ *that commutes with the* L_0*-reduct (in particular, set) forgetful functor.*

[4]When dealing with indices, we write i for K_i.

In this way, our extension of Mal'cev's conception of rational equivalence [15] to algebraic systems is indeed proper.

1.4 Subprevarieties versus extensions

First of all, we have the following immediate consequence of Theorem 1.6, (1.14), (1.15) and Corollaries 1.15, 1.24 and 1.25 of [34]:

Theorem 1.6. *Let* P *be a prevariety of* L*-structures. Then, the mappings* $\mathbb{E} \mapsto \mathrm{Mod}_{\approx}(\mathbb{E})$ *and* $\mathsf{S} \mapsto \vdash^{\approx}_{\mathsf{S}}$ *are inverse to one another dual isomorphisms between the lattice of all extensions of* $\vdash^{\approx}_{\mathsf{P}}$ *and that of all subprevarieties of* P*, corresponding extensions and subprevarieties having the same relative axiomatizations, in which case axiomatic extensions of* $\vdash^{\approx}_{\mathsf{P}}$ *correspond to relative subvarieties of* P*. In particular,* P *is structurally complete iff* $\vdash^{\approx}_{\mathsf{P}}$ *is.*

Theorems 1.6, 1.4 and 1.3 immediately yield:

Corollary 1.7. *Let* P_i*, where* $i \in 3 \setminus 1$*, is a prevariety of* L_i*-structures. Assume* P_1 *and* P_2 *are rationally equivalent with respect to an* $(L_1, L_2 + \approx)$*-translation* $\tau_{1,2}$ *and an* $(L_2, L_1 + \approx)$*-translation* $\tau_{2,1}$*. Then, the mappings* $(\tau_{1,2})^{-1} \circ \mho_2$ *and* $(\tau_{2,1})^{-1} \circ \mho_1$ *are inverse to one another isomorphisms between the lattices of subprevarieties of* P_1 *and* P_2*, corresponding subprevarieties being rationally equivalent with respect to translations involved. Moreover, if a subprevariety* S *of* P_i*, where* $i \in 3 \setminus 1$*, is relatively axiomatized by* $\mathbb{R}$*, then* $(\tau_{i,3-i})^{-1}[\mho_{3-i}[\mathsf{S}]]$ *is relatively axiomatized by* $\tau_{i,3-i}[\mathbb{R}]$*. In particular, relative subvarieties of* P_1 *correspond to those of* P_2*, while* P_1 *is structurally complete iff* P_2 *is.*

1.5 Equivalent algebraic semantics

A class K of L_2-structures [F_2-algebras, in which case $R_2 = \varnothing$] is referred to as a $\langle\kappa$*-ary*$\rangle$ $\{L_0$*-conservatively*$\}$ *equivalent [purely-]algebraic semantics for* a universal Horn L_1-theory $\mathbb{T}$ *(with respect to* $\{L_0$-conservative$\}$ $(L_2 + \approx, L_1)$-translation ρ *and* $(L_1, L_2 + \approx)$-translation τ*)*, provided $\vdash^{\approx}_{\mathsf{K}}$ is $\langle\kappa$-compact and$\rangle$ equivalent to $\mathbb{T}$ (with respect to ρ and τ), in which case $\mathbf{I}(\mathsf{K})$ is so. (We say "relationally" ["functionally"] for "L_0-conservatively" in case $F_1 = F_0 = F_2$ [$R_1 = R_0 = R_2$]. In addition, as usual, we say "finitary" for "ω-ary".) (Recall that, according to Corollary 1.25 of [34], $\vdash^{\approx}_{\mathsf{K}}$ is κ-compact iff the prevariety $\mathbf{I}(\mathsf{K})$ is κ-ary.)

Combining Theorems 1.6 and 1.3, we immediately get:

Corollary 1.8. *Let* P *be a prevariety of* L_2*-structures and* $\mathbb{T}$ *a universal Horn* L_1*-theory. Assume* P *is an equivalent algebraic semantics for* $\mathbb{T}$ *with respect to* ρ *and* τ*. Then, the mappings* $\mathbb{E} \mapsto \mathrm{Mod}_{\approx}(\rho^{-1}[\mathbb{E}])$ *and* $\mathsf{S} \mapsto \tau^{-1}[\vdash^{\approx}_{\mathsf{S}}]$ *are inverse to one another dual isomorphisms between the lattice of all extensions of* $\mathbb{T}$ *and that of all subprevarieties of* P*. Moreover, if an extension* $\mathbb{E}$ *of* $\mathbb{T}$ *is relatively axiomatized by* $\mathbb{R}$*, then the corresponding subprevariety of* P *is relatively axiomatized by* $\tau[\mathbb{R}]$ *and is a relationally equivalent algebraic semantics for* $\mathbb{E}$ *with respect to* ρ *and* τ*. Conversely, if a subprevariety* S *of* P *is relatively axiomatized by* $\mathbb{R}$*, then the corresponding extension of* $\mathbb{T}$ *is relatively axiomatized by* $\rho[\mathbb{R}]$*. In particular, axiomatic extensions of* $\mathbb{T}$ *correspond to relative subvarieties of* P*, while* $\mathbb{T}$ *is structurally complete iff* P *is.*

This extends appropriate results of [24] (cf. Remark 1.1).

1.6 Applications to equivalential universal Horn theories

By $\mathbb{E}_L$ we denote the *equational L-theory*, that is, the finitary universal Horn $(L + \approx)$-theory constituted by the following quasi-identities:

$$\begin{aligned} x &\approx x, \\ (x \approx y) &\rightarrow (y \approx x), \\ (x \approx y)\&(y \approx z) &\rightarrow (x \approx z), \\ (y \approx z) \rightarrow ((f(\overline{x}(\mu_L(f)))[x_i/y]) &\approx (f(\overline{x}(\mu_L(f)))[x_i/z])), \\ (y \approx z)\&(r(\overline{x}(\mu_L(r)))[x_j/y]) &\rightarrow (r(\overline{x}(\mu_L(r)))[x_j/z]), \end{aligned}$$

where $f \in F, i \in \mu_L(f)$ and $r \in R, j \in \mu_L(r)$.

An *equational L-system* is any L-conservative (in which case it is both purely relational and parameter-less) $(L+\approx, L)$-translation ε. It is entirely determined by and, for this reason, identified with the set $\varepsilon(\approx) \subseteq \mathrm{At}_L(V^2)$ alone. In that case, $\mathbb{E}_\varepsilon$ is constituted by the following L-implications:

$$\varepsilon(\approx) \rightarrow (x_0 \approx x_1), \tag{1.4}$$

$$(x_0 \approx x_1) \rightarrow \Phi, \tag{1.5}$$

where $\Phi \in \varepsilon(\approx)$. Then, given any L-structure $\mathcal{A}$, put $\varepsilon^{\mathcal{A}} \triangleq \{\overline{a} \in A^2 | \mathcal{A} \models \varepsilon[x_i/a_i]_{i\in 2}$. Next, ε is referred to as an *equivalence system for* a universal Horn L-theory $\mathbb{T}$, provided $\varepsilon[\mathbb{E}_L] \subseteq \vdash_{\mathbb{T}}$. Finally, $\mathbb{T}$ is said to be *(κ-)equivalential*, whenever it has a (κ-ary) equivalence system.

First of all, we prove the following basic characterization of equivalential UHTs:

Proposition 1.9. *Let $\mathbb{T}$ be a (κ-ary) UHT. Then, the following are equivalent:*

(i) $\mathbb{T}$ *is (κ-)equivalential;*

(ii) $\mathrm{Mod}_{\Im}(\mathbb{T})$ *is a (κ-ary) relationally equivalent algebraic semantics for $\mathbb{T}$;*

(iii) $\mathbb{T}$ *has a (κ-ary) relationally equivalent algebraic semantics;*

(iv) $\mathbb{T}$ *has a (κ-ary) equivalent algebraic semantics.*

Proof. First, (i)$\Rightarrow$(ii) is by Lemma 3.4, Corollary 3.5 and Note 3.6 of [34]. Next, (ii)$\Rightarrow$(iii)$\Rightarrow$(iv) are trivial. Finally, (iv)$\Rightarrow$(i) is by Proposition 2.3(i), (1.14) and Lemma 3.8 of [34]. ■

Recall that, by definition (cf. Section 3.2 of [34]), an UHT is (κ-)algebraizable iff it has a (κ-ary) equivalent purely-algebraic semantics. Thus, the notion of equivalential UHT is the extension of the one of algebraizable UHT resulted from involving algebraic systems instead of pure algebras. Next, we also prove the following analogue of Lemma 3.17 of [34].

Lemma 1.10. *A prevariety P of L_2-structures is an F_0-conservatively equivalent algebraic semantics for a universal Horn L_1-theory $\mathbb{T}$ iff $\mathrm{Mod}_{\Im}(\mathbb{T})$ is a relationally equivalent algebraic semantics for $\mathbb{T}$ F_0-conservatively rationally equivalent to P.*

Proof. The "if" part is due to Proposition 2.9 of [34] and Theorem 1.4. Conversely, assume $\vdash^{\approx}_{\mathsf{P}}$ is F_0-conservatively equivalent to $\mathbb{T}$. Then, by Proposition 1.9, $\mathbb{T}$ is equivalential. Hence, by Lemma 3.4 and Corollary 3.5 of [34], $\vdash^{\approx}_{\mathrm{Mod}_{\Im}(\mathbb{T})}$ is relationally (and so F_0-conservatively) equivalent to $\mathbb{T}$. Therefore, by Proposition 2.9 of [34], $\vdash^{\approx}_{\mathrm{Mod}_{\Im}(\mathbb{T})}$ and $\vdash^{\approx}_{\mathsf{P}}$ are F_0-conservatively equivalent. Thus, by Theorem 1.4, P and $\mathrm{Mod}_{\Im}(\mathbb{T})$ are F_0-conservatively rationally equivalent, as required. ■

Thus, any equivalential UHT has a unique (up to rational equivalence) equivalent algebraic semantics, namely, the one constituted by all simple models of it. This is similar to Lemma 3.17 of [34], according to which any algebraizable

UHT has a unique (up to functional rational equivalence) equivalent purely-algebraic semantics, namely, the one constituted by all underlying algebras of simple models of it. [5]

By Proposition 2.9 of [34] together with Lemma 1.10 and Theorem 1.4, we immediately obtain:

Theorem 1.11. *Let* $\mathbb{T}_i$*, where* $i \in 3 \setminus 1$*, be an equivalential universal Horn* L_i*-theory. Then, the following are equivalent:*

(i) $\mathbb{T}_1$ *and* $\mathbb{T}_2$ *are* L_0*-conservatively equivalent;*

(ii) $\mathrm{Mod}_\Im(\mathbb{T}_1)$ *and* $\mathrm{Mod}_\Im(\mathbb{T}_2)$ *are* L_0*-conservatively rationally equivalent;*

(iii) $\mathbb{T}_1$ *and* $\mathbb{T}_2$ *have same* L_0*-conservatively equivalent algebraic semantics.*

Concluding this section, we present useful results dealing with particular equational systems.

Lemma 1.12. *Let* $\mathbb{T}$ *be a universal Horn* L*-theory and* ε *an equational* L*-system. Then, the following are equivalent:*

(i) ε *is an equivalence system for* $\mathbb{T}$*;*

(ii) $\mathrm{Mod}_\Im(\mathbb{T})$ *is an equivalent algebraic semantics for* $\mathbb{T}$ *with respect to* ε *and* ι_L*;*

(iii) *There is a class of* L'*-structures* K *being an equivalent algebraic semantics for* $\mathbb{T}$ *with respect to some* $(L' + \approx, L)$*-translation* ρ *and some* $(L, L' + \approx)$*-translation* τ *such that* $\rho(\approx) = \varepsilon(\approx)$*.*

Proof. (i)⇒(ii) is by Lemma 3.4 and Corollary 3.5 of [34]. (ii)⇒(iii) is trivial. Finally,(iii)⇒(i) is by (1.14) and Lemma 3.8 of [34]. ■

Proposition 1.13. *Let* $\mathbb{T}$ *be a universal Horn* L*-theory and* ε *an equational* L*-system. Then, for any class of* L*-structures* K *being an equivalent algebraic semantics for* $\mathbb{T}$ *with respect to* ε *and any* $(L, L+\approx)$*-translation* τ*, it holds that* $\mathbf{I}(\mathsf{K}) = \mathrm{Mod}_\Im(\mathbb{T})$*.*

[5] Among other things, this clarifies a one more meaning of the generalized equivalence of UHTs, proposed and studied in [34], because without involving functional translations, the similarity between equivalential and algebraizable UHTs mentioned above could not be manifested at all.

Proof. By Lemma 1.12, ε is an equivalence system for $\mathbb{T}$, whereas $\vdash^{\approx}_{\mathrm{Mod}_{\Im}(\mathbb{T})} = \varepsilon^{-1}[\vdash_{\mathbb{T}}] = \vdash^{\approx}_{\mathsf{K}} = \vdash^{\approx}_{\mathbf{I}(\mathsf{K})}$. Then, by Theorem 3.12(i)⇒(ii) of [34], $\mathrm{Mod}_{\Im}(\mathbb{T})$ is a prevariety, in which case, by (1.15) of [34], we eventually obtain $\mathbf{I}(\mathsf{K}) = \mathrm{Mod}_{\approx}(\vdash^{\approx}_{\mathbf{I}(\mathsf{K})}) = \mathrm{Mod}_{\approx}(\vdash^{\approx}_{\mathrm{Mod}_{\Im}(\mathbb{T})}) = \mathrm{Mod}_{\Im}(\mathbb{T})$, as required. ■

Corollary 1.14. *Let $\mathbb{T}$ be a universal Horn L-theory, ε an equational L-system and $\mathsf{K} \subseteq \mathrm{Mod}_{\approx}(\mathbb{E}_{\varepsilon})$ such that $\vdash_{\mathbb{T}} = \vdash_{\mathsf{K}}$. Then, $\mathbf{I}(\mathsf{K}) = \mathrm{Mod}_{\Im}(\mathbb{T})$.*

Proof. Consider any $\mathcal{A} \in \mathsf{K}$. Then, by (1.4) and (1.5), $\varepsilon^{-1}[\mathcal{A}] = \mathcal{A} + \approx$. Moreover, $(\iota_L)^{-1}[\mathcal{A} + \approx] = \mathcal{A}$. Thus, ε^{-1} and $(\iota_L)^{-1}$ are inverse to one another bijections between K and $\mathsf{K} + \approx$. Hence, by Theorem 2.10(v)⇒(i) of [34], K is an equivalent algebraic semantics for $\mathbb{T}$ with respect to ε and ι_L. In this way, Proposition 1.13 completes the argument. ■

Finally, we prove the following useful characterization of generated prevarieties of simple models of an UHT with equivalence system ε, once, by Lemma 3.1(i)⇒(iv) of [34], the implications in $\mathbb{E}_{\varepsilon}$ are true in those models of the UHT.

Proposition 1.15. *Let ε be an equational L-system and $\mathsf{K} \cup \{\mathcal{A}\} \subseteq \mathrm{Mod}_{\approx}(\mathbb{E}_{\varepsilon})$. Then, $\mathcal{A} \in \mathbf{I}(\mathsf{K})$ iff, for each $r \in R$, $\bigcap\{h^{-1}[r^{\mathcal{B}}] | h \in \mathrm{Hom}(\mathcal{A}, \mathcal{B}), \mathcal{B} \in \mathsf{K}\} = r^{\mathcal{A}}$.*

Proof. The "only if" part is trivial. For proving the "if" part, assume, for each $r \in R$, $\bigcap\{h^{-1}[r^{\mathcal{B}}] | h \in \mathrm{Hom}(\mathcal{A}, \mathcal{B}), \mathcal{B} \in \mathsf{K}\} = r^{\mathcal{A}}$. Take any $a, b \in A$ such that $a \neq b$. Then, by (1.4), there is some $\Phi \in \varepsilon(\approx)$ such that $\mathcal{A} \not\models \Phi[x_0/a, x_1/b]$. On the other hand, $\Phi = r(\overline{\phi})$, for some $r \in R$ and $\overline{\phi} \in \mathrm{Tm}_L(V^2)^{\mu_L(r)}$, in which case $\overline{\phi}^{\mathfrak{A}}[x_0/a, x_1/b] \notin r^{\mathcal{A}}$. Hence, by the assumption, there are some $\mathcal{B} \in \mathsf{K}$ and $h \in \mathrm{Hom}(\mathcal{A}, \mathcal{B})$ such that $\overline{\phi}^{\mathfrak{A}}[x_0/a, x_0/b] \notin h^{-1}[r^{\mathcal{B}}]$, and so $\overline{\phi}^{\mathfrak{B}}[x_0/ha, x_1/hb] \notin r^{\mathcal{B}}$, that is, $\mathcal{B} \not\models \Phi[x_0/ha, x_1/hb]$, in which case, by (1.5), $ha \neq hb$. Thus, $\bigcap\{\ker h | h \in \mathrm{Hom}(\mathcal{A}, \mathcal{B}), \mathcal{B} \in \mathsf{K}\} = \Delta_A$, so $\mathcal{A} \in \mathbf{I}(\mathsf{K})$, as required. ■

Lemma 1.16. *Let ε be an equational L-system and $\mathcal{B}$ an L-structure such that $\varepsilon^{\mathcal{B}} \in \mathrm{Con}(\mathcal{B})$. Then, $\varepsilon^{\mathcal{B}} = \Im(\mathcal{B})$.*

Proof. Consider any $\theta \in \mathrm{Con}(\mathcal{B})$. Take an arbitrary $\langle a, b\rangle \in \theta$. Then, by Corollary 1.3 of [34], we have $\langle a, a\rangle \in \varepsilon^{\mathcal{B}} \Leftrightarrow \langle a, b\rangle \in \varepsilon^{\mathcal{A}}$. Hence, since $\varepsilon^{\mathcal{B}} \in \mathrm{Con}(\mathcal{B})$ is reflexive, we get $\langle a, b\rangle \in \varepsilon^{\mathcal{B}}$. Thus, $\theta \subseteq \varepsilon^{\mathcal{B}}$. Then, Proposition 1.4 of [34] completes the argument. ■

Proposition 1.17. *Let ε be an equational L-system. Then, any member of $\mathrm{Mod}_{\approx}(\mathbb{E}_{\epsilon})$ is simple.*

Proof. Take any $\mathcal{B} \in \mathrm{Mod}_{\approx}(\mathbb{E}_{\epsilon})$. Then, $\varepsilon^{\mathcal{B}} = \Delta_B \in \mathrm{Con}(\mathcal{B})$, in which case Lemma 1.16 completes the argument. ■

Chapter 2

Sequent calculi for finitely-valued logics with equality determinant

During this chapter, we entirely follow conventions adopted in Section 1.1 and Appendix C of [35] except that α (like β) is supposed to be an arbitrary element of $\{\omega, \omega \setminus 1, 2, 2 \setminus 1\}$, in which case all the notions, defined therein for the case $\alpha \in \{\omega, \omega \setminus 1\}$, are extended to the case $\alpha \in \{2, 2 \setminus 1\}$ immediately. As for sentential languages and related issues, we follow the standard conventions, adopted in both [34] and [35].

Fix a sentential language F. Fix also a finite non-singular F-matrix $\mathcal{A}$ with finite *equality determinant* Υ (cf. [31]), that is, $x_0 \in \Upsilon \subseteq \mathrm{Tm}_F(V^1)$, while, for all $a, b \in A$, it holds that $a = b \Leftrightarrow \forall \phi \in \Upsilon : \phi^{\mathfrak{A}}[x_0/a] \in D^{\mathcal{A}} \Leftrightarrow \phi^{\mathfrak{A}}[x_0/b] \in D^{\mathcal{A}}$, in which case $\varnothing \neq D^{\mathcal{A}} \neq A$, for, otherwise, $\mathcal{A}$ would be singular.

A 0-(1-)*subalgebra of* $\mathcal{A}$ is any subalgebra $\mathfrak{B}$ of $\mathfrak{A}$ such that $B \cap D^{\mathcal{A}} = \varnothing$ ($B \subseteq D^{\mathcal{A}}$) (cf. [31]). For any $i \in 2$, put:

$$\gamma_i^{\mathcal{A}} \triangleq \begin{cases} 1 & \text{if } \mathcal{A} \text{ has an } i\text{-subalgebra,} \\ 0 & \text{otherwise.} \end{cases}$$

The work [31] suggestes an F-sequent calculus $\mathbb{G}$ of type $[\omega \setminus \gamma_0^{\mathcal{A}}, \omega \setminus \gamma_1^{\mathcal{A}}]$ with structural rules such that:

$$\vdash_{\mathbb{G}} = \vdash_{\mathcal{A}^{[\omega \setminus \gamma_0^{\mathcal{A}}, \omega \setminus \gamma_1^{\mathcal{A}}]}}. \tag{2.1}$$

Put $\varepsilon_\Upsilon \triangleq \{\phi[x_0/x_i] \rightarrowtail (\phi[x_0/x_{1-i}]) | \phi \in \Upsilon, i \in 2\}$. Then, as Υ is an equality determinant for $\mathcal{A}$, while $1 \in \alpha \cap \beta$, we have (cf. (C.1) of [35]):

$$(\varepsilon_\Upsilon)^{\mathcal{A}^{[\alpha,\beta]}} = \Delta_A. \tag{2.2}$$

In this way, by Lemma 3.1(v)⇒(i) of [34], (2.1), (2.2) and Corollary 1.14, we get:

Theorem 2.1. (cf. Proposition 10 and Lemma 6 of [32]) ε_Υ *is a finitary equivalence system for* $\mathbb{G}$*, while* $\mathrm{Mod}_\Im(\mathbb{G}) = \mathbf{I}(\mathcal{A}^{[\omega\setminus\gamma_0^{\mathcal{A}},\omega\setminus\gamma_1^{\mathcal{A}}]})$.

Thus, in view of Corollary 1.8 and Lemma 1.12(i)⇒(ii), to study extensions of $\mathbb{G}$ is to study subprevarieties of $\mathbf{I}(\mathcal{A}^{[\omega\setminus\gamma_0^{\mathcal{A}},\omega\setminus\gamma_1^{\mathcal{A}}]})$. On the other hand, $F^{[\omega\setminus\gamma_0^{\mathcal{A}},\omega\setminus\gamma_1^{\mathcal{A}}]}$-structures are too cumbersome to deal with them directly. Nevertheless, the primary structure $\mathcal{A}^{[\omega\setminus\gamma_0^{\mathcal{A}},\omega\setminus\gamma_1^{\mathcal{A}}]}$ admits a much more compact relationally rationally equivalent representation. We start from providing an intermediate one of the signature $F^{[\omega\setminus 1,\omega\setminus 1]}$.

An 0-(1-)*tautology of* $\mathcal{A}$ is any $\overline{\phi} \in \mathrm{Tm}_F(V^1)^m$, where $m \in \omega\setminus 1$, such that $\varnothing \vdash_{\mathfrak{A}^{[\omega,\omega]}} (\rightarrowtail \overline{\phi})$ (resp., $\vdash_{\mathfrak{A}^{[\omega,\omega]}} (\overline{\phi} \rightarrowtail)$).

Remark 2.2. (cf. Appendix C) Any 0-subalgebra of $\mathcal{A}$ is a 1-subalgebra of the complimentary matrix $\overline{\mathcal{A}} = \langle \mathfrak{A}, A\setminus D^{\mathcal{A}}\rangle$, Moreover, any 0-tautology of $\mathcal{A}$ is a 1-tautology of $\overline{\mathcal{A}}$. ∎

Proposition 2.3. *For each* $i \in 2$*,* $\mathcal{A}$ *has an* i*-tautology, whenever it has no* i*-subalgebra.*

Proof. Let $i = 0$. Take any $\overline{a} \in (A\setminus D^{\mathcal{A}})^n$, where $n \triangleq |A\setminus D^{\mathcal{A}}| \in \omega\setminus 1$. Then, for each $j \in n$, there is some $\phi_j \in \mathrm{Tm}_L(V^1)$ such that $\phi_j^{\mathfrak{A}}[x_0/a_j] \in D^{\mathcal{A}}$. In that case, $(\overline{\phi}, x_0)$ is a 0-tautology of $\mathcal{A}$.

In view of Remark 2.2 and (C.1), the case $i = 0$ equally yields the case $i = 1$. ∎

According to Proposition 2.3, from now on, for each $i \in 2$, fix an arbitrary i-tautology $\overline{\phi}_i$ of $\mathcal{A}$, whenever $\mathcal{A}$ has no i-subalgebra, in which case:

$$\varnothing \vdash_{\mathcal{A}^{[\omega,\omega]}} \{\varphi \rightarrowtail \overline{\phi}_0 | \varphi \in \mathrm{img}\,\overline{\phi}_0[x_0/x_1]\}, \tag{2.3}$$

$$\varnothing \vdash_{\mathcal{A}^{[\omega,\omega]}} \{\overline{\phi}_1 \rightarrowtail \psi | \psi \in \mathrm{img}\,\overline{\phi}_1[x_0/x_1]\}. \tag{2.4}$$

Define the purely-relational $(F^{[\alpha,\beta]}, F^{[\alpha\setminus(1-\gamma_0^{\mathcal{A}}),\beta]}$-translation $v_0^{[\alpha,\beta]}$ as the extension of $\iota_{F^{[\alpha\setminus(1-\gamma_0^{\mathcal{A}}),\beta]}}$ to $F^{[\alpha,\beta]}$ given by (where $n \in \beta$):

$$v_0^{[\alpha,\beta]}(\rightarrowtail_n^0) \triangleq \{\varphi \rightarrowtail \overline{x}(n) | \varphi \in \mathrm{img}\,\overline{\phi}_0\},$$

whenever $\mathcal{A}$ has no 0-subalgebra, while $0 \in \alpha$. Clearly, we have, the compatibility of the structure with the translation being by (2.3):

$$\mathcal{A}^{[\alpha,\beta]} = (v_0^{[\alpha,\beta]})^{-1}[\mathcal{A}^{[\alpha\setminus(1-\gamma_0^{\mathcal{A}}),\beta]}]. \tag{2.5}$$

Likewise, define the purely-relational $(F^{[\alpha,\beta]}, F^{[\alpha,\beta\backslash(1-\gamma_1^A)]}$-translation $\upsilon_1^{[\alpha,\beta]}$ as the extension of $\iota_{F^{[\alpha,\beta\backslash(1-\gamma_1^A)]}}$ to $F^{[\alpha,\beta]}$ given by (where $m \in \alpha$):

$$\upsilon_1^{[\alpha,\beta]}(\rightarrowtail_0^m) \triangleq \{\overline{x}(m) \rightarrowtail \psi | \psi \in \operatorname{img} \overline{\phi}_1\},$$

whenever $\mathcal{A}$ has no 1-subalgebra, while $0 \in \beta$. Then, we have, the compatibility of the structure with the translation being by (2.4):

$$\mathcal{A}^{[\alpha,\beta]} = (\upsilon_1^{[\alpha,\beta]})^{-1}[\mathcal{A}^{[\alpha,\beta\backslash(1-\gamma_1^A)]}]. \tag{2.6}$$

In this way, we have the purely relational $(F^{[\alpha,\beta]}, F^{[\alpha\backslash(1-\gamma_0^A),\beta\backslash(1-\gamma_1^A)]}$-translation:[1]

$$\upsilon^{[\alpha,\beta]} \triangleq \upsilon_1^{[\alpha\backslash(1-\gamma_0^A),\beta]} \circ \upsilon_0^{[\alpha,\beta]}.$$

Then, by (2.5) and (2.6), we get, the compatibility of the structure with the translation being by both (2.3) and (2.4):

$$\mathcal{A}^{[\alpha,\beta]} = (\upsilon^{[\alpha,\beta]})^{-1}[\mathcal{A}^{[\alpha\backslash(1-\gamma_0^A),\beta\backslash(1-\gamma_1^A)]}]. \tag{2.7}$$

2.1 Logics with truth and falsehood constants

During this section, it is supposed that F contains both "truth" $\top$ and "falsehood" $\bot$ constants. Assume also that $\top^{\mathfrak{A}} \in D^{\mathcal{A}}$, while $\bot^{\mathfrak{A}} \notin D^{\mathcal{A}}$. In that case, $\mathcal{A}$ has neither 0- nor 1-subalgebra while $\top$ and $\bot$ are variable-free 0- and 1-tautology, respectively. Moreover,

$$\upsilon_0^{[\alpha,\beta]}(\rightarrowtail_n^0) = \{\top \rightarrowtail \overline{x}(n)\}, \tag{2.8}$$

where $n \in \beta$, while $0 \in \alpha$. Likewise,

$$\upsilon_1^{[\alpha,\beta]}(\rightarrowtail_0^m) \triangleq \{\overline{x}(m) \rightarrowtail \bot\}, \tag{2.9}$$

where $m \in \alpha$, while $0 \in \beta$.

[1]As a composition of purely-relational translations is unique, we use the standard notation for composition in such cases.

2.2 Logics with conjunction and disjunction

Throughout the rest of this chapter as well during the next one, it is supposed that F contains binary connectives $\wedge$ and $\vee$ (*conjunction* and *disjunction*, respectively). To unify further notations, while F-algebras are denoted by Fraktur letters $\mathfrak{A}, \mathfrak{B}, \mathfrak{C}, \ldots$ (possibly with indices), $\{\wedge, \vee\}$-algebras [being reducts of the former ones] are denoted by [corresponding] Script letters $\mathcal{A}, \mathcal{B}, \mathcal{C}, \ldots$ (with [the same] indices if any), their carriers being denoted by corresponding capital Italic letters $A, B, C, \ldots$ (with respective indices if any). Given an F-algebra $\mathfrak{B}$, a subset $X \subseteq B$ is referred to as a *(prime) filter of* $\mathfrak{B}$, whenever, for all $a, b \in B$, it holds that $\{a, b\} \subseteq X \Leftrightarrow a \wedge^{\mathfrak{B}} b \in X$ (resp., and $\{a, b\} \cap X \neq \varnothing \Leftrightarrow a \vee^{\mathfrak{B}} b \in X$).

From now on, it is also assumed that $D^{\mathcal{A}}$ is a prime filter of $\mathcal{A}$ (cf. [32]), in which case $\mathcal{A}$ is both $\{x_0 \wedge x_1\}$-conjunctive and $\{x_0 \vee x_1\}$-disjunctive, while the binary F-systems $\{x_0 \wedge x_1\}$ and $\{x_0 \vee x_1\}$ are both parameter-less.

Consider the purely relational parameter-less $(F^{[\omega\setminus 1, \omega\setminus 1]}, F^{[\{1\},\{1\}]})$-translation

$$\chi^{\wedge}_{\vee} \triangleq \chi^{\{x_0 \wedge x_1\}} \circ \chi_{\{x_0 \vee x_1\}},$$

in which case (where $m, n \in \omega \setminus 1$):

$$\chi^{\wedge}_{\vee}(\rightarrowtail^{m}_{n}) = \{\wedge \bar{x}(m) \rightarrowtail \vee \bar{x}(m, n)\}. \tag{2.10}$$

As $D^{\mathcal{A}}$ is a prime filter of $\mathcal{A}$, we obviously have:

$$\mathcal{A}^{[\omega\setminus 1, \omega\setminus 1]} = (\chi^{\wedge}_{\vee})^{-1}[\mathcal{A}^{[\{1\},\{1\}]}]. \tag{2.11}$$

Combining (2.7) and (2.11) with Theorem 1.4(iv)⇒(iii), we get, the compatibility of the structure with the translation being by both (2.3) and (2.4):

Proposition 2.4. *$\mathcal{A}^{[\{1\},\{1\}]}$ and $\mathcal{A}^{[\omega\setminus\gamma^{\mathcal{A}}_0, \omega\setminus\gamma^{\mathcal{A}}_1]}$ are relationally rationally equivalent with respect to $\iota_{F^{[\{1\},\{1\}]}}$ and $\chi^{\wedge}_{\vee} \circ \upsilon^{[\omega\setminus\gamma^{\mathcal{A}}_0, \omega\setminus\gamma^{\mathcal{A}}_1]}$, and so are $\mathbf{I}(\mathcal{A}^{[\{1\},\{1\}]})$ and $\mathbf{I}(\mathcal{A}^{[\omega\setminus\gamma^{\mathcal{A}}_0, \omega\setminus\gamma^{\mathcal{A}}_1]})$.*

In case $\mathcal{A}$ is, in addition, as in Section 2.1, by (2.8), (2.9), (2.10), we have:

$$(\chi^{\wedge}_{\vee} \circ \upsilon^{[\omega,\omega]})(\rightarrowtail^{m}_{n}) = \{\wedge(\bar{x}(m), \top) \rightarrowtail \vee(\bar{x}(m, n), \bot)\},$$

where $m, n \in \omega$.

Thus, in view of Corollary 1.7, to study subprevarieties of $\mathbf{I}(\mathcal{A}^{[\omega\setminus\gamma^{\mathcal{A}}_0, \omega\setminus\gamma^{\mathcal{A}}_1]})$ is to study those of $\mathbf{I}(\mathcal{A}^{[\{1\},\{1\}]})$. The latter is easier than the former, once the signature $F^{[\{1\},\{1\}]}$ contains just a single (binary) relation symbol $\rightarrowtail^1_1$, hereafter denoted by Q to simplify further notations.

Before pursuing, we should like to highlight that Appendices provide all necessary tools for proper constructing various equivalent fragmentary versions (in particular, those of types $[\omega \setminus 1, \omega \setminus 1]$ and $[\{1\}, \{1\}]$) of the initial calculus $\mathbb{G}$. However, we refrain from specifying them explicitly here, because this point being mainly of proof-theoretical nature is no matter for our further purely model-theoretic discussion.

2.2.1 Quasi-ordered algebras

Put $L \triangleq \langle F, \{Q\}\rangle$. Any L-structure $\mathcal{B}$ is identified with the couple $\langle \mathfrak{B}, Q^{\mathcal{B}}\rangle$.

The variety of all F-algebras satisfyng the lattice identities (cf. [2]) is denoted by L_F. The partial ordering of any $\mathfrak{B} \in \mathsf{L}_F$ is denoted by $\leqslant^{\mathfrak{B}}$.

Clearly, $\mathcal{A}^{[\{1\},\{1\}]} = \langle \mathfrak{A}, \varpi_A^{D^{\mathcal{A}}}\rangle$. Then, since $\varpi_A^{D^{\mathcal{A}}}$ is a quasi-ordering on A, while $D^{\mathcal{A}}$ is a prime filter of $\mathcal{A}$, the following quasi-identities are true in $\mathcal{A}^{[\{1\},\{1\}]}$, and so in $\mathbf{I}(\mathcal{A}^{[\{1\},\{1\}]})$:[2]

$$xQx, \tag{2.12}$$
$$(xQy)\&(yQz) \rightarrow (xQz), \tag{2.13}$$
$$xQ(x \vee y), \tag{2.14}$$
$$(xQz)\&(yQz) \rightarrow (x \vee y)Qz, \tag{2.15}$$
$$(x \wedge y)Qx, \tag{2.16}$$
$$(zQx)\&(zQy) \rightarrow zQ(x \wedge y). \tag{2.17}$$

A *quasi-ordered F-algebra* is any L-structure satisfying the quasi-identities (2.12)—(2.17) and the lattice identities.[3]

Put $\varepsilon \triangleq \varepsilon_\Upsilon$.

Remark 2.5. In view of (2.12), the quasi-identities (1.5) are all true in quasi-ordered F-algebras. ∎

Remark 2.6. In view of (2.12), any singular quasi-ordered F-algebra is trivial. ∎

Given any quasi-ordered F-algebra $\mathcal{B}$, we have the equivalence $E^{\mathcal{B}} \triangleq Q^{\mathcal{B}} \cup (Q^{\mathcal{B}})^{-1}$ on B as well as the corresponding natural mapping $\nu^{\mathcal{B}} \triangleq \nu_{E^{\mathcal{B}}}$.

---

[2]We use the natural infix notation for the binary relation symbol Q.

[3]As a matter of fact, for all the particular cases of $\mathcal{A}$, studied in Chapter 3, $\mathcal{A}$ is a lattice, so this definition of the concept of quasi-ordered algebra is well-justified for our purpose.

Proposition 2.7. *Let $\mathcal{B}$ be a quasi-ordered F-algebra. Then,* $\mathrm{Con}(\mathcal{B}) = \{\theta \in \mathrm{Con}(\mathfrak{B}) | \theta \subseteq Q^{\mathcal{B}}\}$.

Proof. Take any $\theta \in \mathrm{Con}(\mathcal{B})$. Then, $\theta \in \mathrm{Con}(\mathfrak{B})$. Moreover, consider any $\langle a, b\rangle \in \theta$. Then, since $aQ^{\mathcal{B}}a$, we get $aQ^{\mathcal{B}}b$. Thus, $\theta \subseteq Q^{\mathcal{B}}$. Conversely, consider any $\theta \in \mathrm{Con}(\mathfrak{B})$ such that $\theta \subseteq Q^{\mathcal{B}}$, in which case $\theta \subseteq E^{\mathcal{B}}$. Take any $a, b, c, d \in B$ such that $\langle a, b\rangle \in Q^{\mathcal{B}}$, while $\langle a, c\rangle, \langle b, d\rangle \in \theta$. Then, we obviously have $\langle c, d\rangle \in Q^{\mathcal{B}}$. Thus, $\theta \in \mathrm{Con}(\mathcal{B})$, as required. ■

Lemma 2.8. *Let $\mathcal{B}$ be a quasi-ordered F-algebra. Then, $E^{\mathcal{B}} \in \mathrm{Con}(\mathcal{B})$,*

Proof. It is easy to check that the following quasi-identities are true in quasi-ordered F-algebras:

$$(x_1 Q y_1)\&(x_2 Q y_2) \rightarrow (x_1 \wedge x_2)Q(y_1 \wedge y_2), \tag{2.18}$$

$$(x_1 Q y_1)\&(x_2 Q y_2) \rightarrow (x_1 \vee x_2)Q(y_1 \vee y_2). \tag{2.19}$$

This completes the argument. ■

Lemma 2.9. *Let $\mathcal{B}$ be a quasi-ordered F-algebra such that $\varepsilon^{\mathcal{B}} \in \mathrm{Con}(\mathfrak{B})$. Then, $\varepsilon^{\mathcal{B}} = \Im(\mathcal{B})$.*

Proof. First, the fact that $\varepsilon^{\mathcal{B}} \subseteq Q^{\mathcal{B}}$ is by $(x_0 Q x_1) \in \varepsilon$, for $x_0 \in \Upsilon$. Then, by Proposition 2.7, $\varepsilon^{\mathcal{B}} \in \mathrm{Con}(\mathcal{B})$. Thus, Lemma 1.16 completes the argument. ■

Put $(x \lessapprox y) \triangleq (x \approx x \wedge y)$. By (2.12) and (2.16), the following quasi-identity is true in quasi-ordered F-algebras:

$$(x \lessapprox y) \rightarrow (xQy). \tag{2.20}$$

In this way, for any quasi-ordered F-algebra $\mathcal{B}$, it holds that $\leqslant^{\mathcal{B}} \subseteq Q^{\mathcal{B}}$.

A *partially-ordered F-algebra* is any quasi-ordered F-algebra satisfying the quasi-identity:

$$(xQy)\&(yQx) \rightarrow (x \approx y), \tag{2.21}$$

in which case it satisfies (1.4), for $x_0 \in \Upsilon$. The quasivariety of all partially-ordered F-algebras is denoted by PA_F.

Lemma 2.10. $\mathsf{PA}_F = \{\langle \mathfrak{B}, \leqslant^{\mathfrak{B}}\rangle | \mathfrak{B} \in \mathsf{L}_F\}$.

Proof. The inclusion from right to left is evident. For proving the converse, consider any $\mathcal{B} \in \mathsf{PA}_F$, in which case $\mathfrak{B} \in \mathsf{L}_F$. Take an arbitrary $\langle a, b\rangle \in Q^{\mathcal{B}}$. Then, by (2.12), (2.16), (2.17), we have $(a \wedge^{\mathfrak{B}} b)E^{\mathcal{B}}a$, so, by (2.21), we get $a \leqslant^{\mathfrak{B}} b$. Thus, $Q^{\mathcal{B}} \subseteq \leqslant^{\mathfrak{B}}$. Then, (2.20) completes the argument. ■

Remark 2.11. In view of Lemma 2.10, the quasivariety PA_F is relationally rationally equivalent to the variety L_F with respect to the purely-relational $(L, F + \approx)$-translation $\tau_{\lessapprox}$, given by $\tau_{\lessapprox}(Q) \triangleq \{x_0 \lessapprox x_1\}$, and ι_F. ■

Remark 2.12. Given any quasi-ordered F-algebra $\mathcal{B}$, $\langle \mathfrak{B}, Q^{\mathcal{B}} \rangle$ is a quasi-ordered $\{\wedge, \vee\}$-algebra, while $E^{\mathcal{B}} \subseteq Q^{\mathcal{B}}$, in which case by Lemma 2.8 and Proposition 2.7, $E^{\mathcal{B}} \in \mathrm{Con}(\langle \mathfrak{B}, Q^{\mathcal{B}} \rangle)$. On the other hand, as $\mathfrak{B}/E^{\mathcal{B}}$ is a lattice, in view of Lemma 1.14 of [34], $\mathcal{C} \triangleq \langle \mathfrak{B}, Q^{\mathcal{B}} \rangle / E^{\mathcal{B}}$ is a quasi-ordered $\{\wedge, \vee\}$-algebra as well. Moreover, it is routine checking that $E^{\mathcal{C}} = \Delta_C$, for $Q^{\mathcal{C}} = \nu^{\mathcal{B}}[Q^{\mathcal{B}}]$, so $\mathcal{C}$ is a partially-ordered $\{\wedge, \vee\}$-algebra, in which case, by Lemma 2.10, $\mathfrak{C} = \langle \mathfrak{B}/E^{\mathcal{B}}, \leqslant^{\mathfrak{B}/E^{\mathcal{B}}} \rangle$, so $\nu^{\mathcal{B}}[Q^{\mathcal{B}}] = \leqslant^{\mathfrak{B}/E^{\mathcal{B}}}$, while $Q^{\mathcal{B}} = (\nu^{\mathcal{B}})^{-1}[\leqslant^{\mathfrak{B}/E^{\mathcal{B}}}]$. ■

Chapter 3

Examples

Here, we explore examples of equivalential but non-algebraizable universal Horn theories. They all belong to the class described in Chapter 2. Moreover, Section 3.3 equally covers the sentential logic of the corresponding matrix with equality determinant for the matrix is relationally rationally equivalent to the corresponding quasi-ordered algebra.

3.1 Implication-less linear intuitionistic logic

Throughout this section, put $F \triangleq \{\wedge, \vee, \neg, \top, \bot\}$, where $\neg$ is unary (negation).

Recall that a *Stone algebra* (cf. [2], [9], [10]) is any F-algebra, whose $(F \setminus \{\neg\})$-reduct is a bounded distributive lattice (cf. [2]) and the following identities are true in it:[1]

$$x \wedge \neg x \approx \bot, \tag{3.1}$$

$$\neg\bot \approx \top, \tag{3.2}$$

$$\neg\top \approx \bot, \tag{3.3}$$

$$x \wedge \neg(x \wedge y) \approx x \wedge \neg y, \tag{3.4}$$

$$\neg x \vee \neg\neg x \approx \top, \tag{3.5}$$

the variety of all them being denoted by SA.

[1]The first four identities (together with the identities axiomatizing the variety of bounded distributive lattices) axiomatize the variety of *pseudo-complemented bounded distributive lattices* (cf. [2]), the third identity being a consequence of the first one, while the fifth identity is a relative axiomatization of the subvariety of Stone algebras (cf. [10]).

A Stone algebra is referred to as a *Boolean algebra*, provided it satisfies the identity:

$$x \vee \neg x \approx \top, \tag{3.6}$$

the variety of all them being denoted by BA. Recall that the following *De Morgan* identities are true in Stone algebras:[2]

$$\neg(x \vee y) \approx \neg x \wedge \neg y, \tag{3.7}$$
$$\neg(x \wedge y) \approx \neg x \vee \neg y, \tag{3.8}$$

In this way, by (3.5), for any Stone algebra $\mathfrak{B}$, the set $\neg^{\mathfrak{B}}[B]$ forms a Boolean subalgebra of $\mathfrak{B}$.

A *quasi-[partially-]ordered Stone (Boolean) algebra* is any quasi-[partially-]ordered F-algebra, whose underlying algebra is a Stone (resp., Boolean) algebra, the quasivariety of all them being denoted by QSA [resp. by PSA] (resp., by QBA).

Lemma 3.1. *Let $\mathcal{B}$ be a quasi-ordered Stone algebra. Then,* $E^{\mathcal{B}} \cap (\neg^{\mathfrak{B}}[B])^2 \in \mathrm{Con}(\mathfrak{B} \upharpoonright (\neg^{\mathfrak{B}}[B]))$,

Proof. By Lemma 2.8, we have $E^{\mathcal{B}} \in \mathrm{Con}(\mathcal{B})$, so $E^{\mathcal{B}} \cap (\neg^{\mathfrak{B}}[B])^2 \in \mathrm{Con}(\mathcal{B} \upharpoonright (\neg^{\mathfrak{B}}[B]))$. Finally, consider any Boolean algebra $\mathfrak{C}$ and any $\theta \in \mathrm{Con}(\mathcal{C})$. Take an arbitrary $\langle a_0, a_1 \rangle \in \theta$. Then, by (3.1) and (3.6), we have, for each $i \in 2$:

$$\begin{aligned}
\neg^{\mathfrak{C}} a_i &= \neg^{\mathfrak{C}} a_i \wedge^{\mathfrak{C}} \top^{\mathfrak{C}} \\
&= \neg^{\mathfrak{C}} a_i \wedge^{\mathfrak{C}} (a_{1-i} \vee^{\mathfrak{C}} \neg^{\mathfrak{C}} a_{1-i}) \\
&\theta\ \neg^{\mathfrak{C}} a_i \wedge^{\mathfrak{C}} (a_i \vee^{\mathfrak{C}} \neg^{\mathfrak{C}} a_{1-i}) \\
&= \neg^{\mathfrak{C}} a_i \wedge^{\mathfrak{C}} \neg^{\mathfrak{C}} a_{1-i},
\end{aligned}$$

so $\langle \neg^{\mathfrak{C}} a_0, \neg^{\mathfrak{C}} a_1 \rangle \in \theta$. Thus, $\theta \in \mathrm{Con}(\mathfrak{C})$. This completes the argument. ∎

Set $\Upsilon \triangleq \{x_0, \neg x_0\}$.

Proposition 3.2. *Let $\mathcal{B}$ be a quasi-ordered Stone algebra. Then,* $\varepsilon^{\mathcal{B}} = \Im(\mathcal{B})$.

Proof. By Lemma 2.8, (3.7) and (3.8), we have $\varepsilon^{\mathcal{B}} \in \mathrm{Con}(\mathcal{B})$. Moreover, by Lemma 3.1, the following quasi-identity is true in QSA:

$$(\neg x Q \neg y) \& (\neg y Q \neg x) \to (\neg\neg x Q \neg\neg y), \tag{3.9}$$

so we get $\varepsilon^{\mathcal{B}} \in \mathrm{Con}(\mathfrak{B})$. Then, Lemma 2.9 completes the argument. ∎

---

[2]The former identity is true in the variety of all pseudo-complemented bounded distributive lattices, the latter being a one more relative axiomatization of the subvariety of Stone algebras (cf. [10]).

Thus, by Remark 2.5 and Proposition 3.2, we see that a quasi-ordered Stone algebra is simple iff it satisfies the quasi-identity (1.4). Hence, the class SQSA (SQBA) of all simple quasi-ordered Stone (resp., Boolean) algebras is a quasi-variety.

Put $\mathcal{S}_3 \triangleq \langle \mathfrak{S}_3, \varpi_3^{3\setminus 2} \rangle$, where $\mathfrak{S}_3$ is the tree-element chain Stone algebra with carrier $S_3 \triangleq 3$ and natural partial ordering given by $0 < 1 < 2$, in which case $\neg^{\mathfrak{S}_3} a \triangleq \max\{b \in 3 | \min(a,b) = 0\}$, for all $a \in 3$. The F-matrix $\mathcal{M}_3 \triangleq \langle \mathfrak{S}_3, 3 \setminus 2 \rangle$ with equality determinant Υ defines the implication-less fragment of Dummett's *linear intuitionistic logic* [7]. Notice that $\mathcal{S}_3 = (\mathcal{M}_3)^{[\{1\},\{1\}]}$.

Lemma 3.3. *Let $\mathfrak{A}$ be a Stone algebra and X a proper prime non-empty filter of $\mathcal{A}$. Then, we have the $h_X^{\mathfrak{A}} \in \mathrm{Hom}(\mathfrak{A}, \mathfrak{S}_3)$ defined by:*

$$h_X^{\mathfrak{A}} a \triangleq \begin{cases} 0 & \textit{if } a \notin X \textit{ and } \neg^{\mathfrak{A}} a \in X, \\ 1 & \textit{if } a \notin X \textit{ and } \neg^{\mathfrak{A}} a \notin X, \\ 2 & \textit{if } a \in X \textit{ and } \neg^{\mathfrak{A}} a \notin X, \end{cases}$$

for all $a \in A$, in which case $X = (h_X^{\mathfrak{A}})^{-1}[3 \setminus 2]$.

Proof. Since X is proper, $\bot^{\mathfrak{A}} \notin F$. Hence, by (3.1), for each $a \in A$, either $a \notin X$ or $\neg^{\mathfrak{A}} a \notin X$. Next, as X is non-empty, $\top^{\mathfrak{A}} \in X$. Then, (3.7), (3.8) and the fact that $3 \setminus 2$ is a proper prime non-empty filter of $\mathfrak{S}_3$ complete the argument. ∎

Theorem 3.4. $\mathsf{SQSA} = \mathbf{I}(\mathcal{S}_3)$.

Proof. Since $\mathfrak{S}_3$ is a Stone algebra, while $3 \setminus 2$ is a prime filter of it, $\mathcal{S}_3$ is a quasi-ordered Stone algebra. Moreover, since Υ is an equality determinant for $\mathcal{M}$, (1.4) is true in $\mathcal{S}_3$, so $\mathcal{S}_3 \in \mathsf{SQSA}$, in which case $\mathbf{I}(\mathcal{S}_3) \subseteq \mathsf{SQSA}$. For proving the converse, consider any $\mathcal{A} \in \mathsf{SQSA}$. Take an arbitrary $\langle a, b \rangle \in A^2 \setminus Q^{\mathcal{A}}$. Then, in view of Remark 2.12, $\nu^{\mathcal{A}}(a) \not\leqslant^{\mathcal{A}/E^{\mathcal{A}}} \nu^{\mathcal{A}}(b)$. Hence, as the quotient $\mathcal{A}/E^{\mathcal{A}}$ is a distributive lattice as well as $\mathcal{A}$ is, by the Prime Ideal Theorem for distributive lattices, there is a prime filter Y of $\mathcal{A}/E^{\mathcal{A}}$ such that $\nu^{\mathcal{A}} a \in Y$, while $\nu^{\mathcal{A}}(b) \notin Y$, in which case $X \triangleq (\nu^{\mathcal{A}})^{-1}[Y]$ is a prime filter of $\mathcal{A}$ such that $a \in X$, while $b \notin X$, so X is both proper and non-empty, whereas $\langle a, b \rangle \notin \varpi_A^X$. Moreover, since $\leqslant^{\mathcal{A}/E^{\mathcal{A}}}[Y] \subseteq Y$, in view of Remark 2.12, we have $Q^{\mathcal{A}}[X] \subseteq X$. and so $Q^{\mathcal{A}} \subseteq \varpi_A^X$. Then, by Lemma 3.3, $h_X^{\mathfrak{A}} \in \mathrm{Hom}(\mathfrak{A}, \mathfrak{S}_3)$, while $X = (h_X^{\mathfrak{A}})^{-1}[3 \setminus 2]$, so $\varpi_A^X = (h_X^{\mathfrak{A}})^{-1}[Q^{\mathcal{S}_3}]$, in which case $\langle a, b \rangle \notin (h_X^{\mathfrak{A}})^{-1}[Q^{\mathcal{S}_3}]$, while $Q^{\mathcal{A}} \subseteq (h_X^{\mathfrak{A}})^{-1}[Q^{\mathcal{S}_3}]$, and so $h_X^{\mathfrak{A}} \in \mathrm{Hom}(\mathcal{A}, \mathcal{S}_3)$. Thus, by Propositions 3.2 and 1.15, $\mathcal{A} \in \mathbf{I}(\mathcal{S}_3)$, as required. ∎

Thus, in view of (1.2) and Theorem 3.4, for analyzing relative subvarieties of SQSA it suffices to analyze those of $\mathbf{S}(\mathcal{S}_3)$. First of all, notice that any $B \subseteq S_3$ forms a subalgebra of $\mathfrak{S}_3$ iff $B \in \{3, \{0, 2\}\}$. Put $\mathcal{S}_2 \triangleq \mathcal{S}_3 \restriction \{0, 2\}$.

The quasivariety of all one-element quasi-ordered Stone algebras is denoted by OQSA. Clearly, OQSA $\cap \mathbf{S}(\mathcal{S}_3) = \varnothing$ (cf. Remark 2.6), while OQSA $= \mathbf{I}(\varnothing) \subseteq$ SQSA.

Note that any $\mathcal{B} \in \mathbf{S}(\mathcal{S}_3)$ satisfies (3.6) iff $B = \{0, 2\}$. Hence, by (1.2) and Theorem 3.4, we have:

Corollary 3.5. SQBA $= \mathbf{I}(\mathcal{S}_2)$.

As any relative subvariety of any class is closed under $\mathbf{S}$, by (1.2) and Theorem 3.4, we eventually get:

Corollary 3.6. *Relative subvarieties of* SQSA *form the three-element chain:*

$$\mathsf{OQSA} \subsetneq \mathsf{SQBA} \subsetneq \mathsf{SQSA}.$$

It only remains to analyze other subprevarieties of SQSA.

A simple quasi-ordered Stone algebra is said to be *pseudo-complete*, provided it satisfies the quasi-identity:

$$(x \vee \neg x) Q \bot \to (\bot \approx \top). \tag{3.10}$$

The quasivariety of all pseudo-complete simple quasi-ordered Stone algebras is denoted by PCSQSA.

Let $\mathcal{A} \in$ QSA. Put:

$$Z^{\mathcal{A}} \triangleq \{a \in A | \exists b \in A : (b \vee^{\mathfrak{A}} \neg^{\mathfrak{A}} b) Q^{\mathcal{A}} a\}.$$

Lemma 3.7. *Let* $\mathcal{A} \in$ QSA. *Then,* $Z^{\mathcal{A}}$ *is a non-empty filter of* $\mathcal{A}$.

Proof. As $\top^{\mathfrak{A}} \vee^{\mathfrak{A}} \neg^{\mathfrak{A}} \top^{\mathfrak{A}} = \top^{\mathfrak{A}}$, by (2.12), $\top^{\mathfrak{A}} \in Z^{\mathcal{A}}$, so $Z^{\mathcal{A}}$ is not empty. Next, take any $c, d \in Z^{\mathcal{A}}$. Then, there are some $a, b \in A$ such that $(a \vee^{\mathfrak{A}} \neg^{\mathfrak{A}} a) Q^{\mathcal{A}} c$ and $(b \vee^{\mathfrak{A}} \neg^{\mathfrak{A}} b) Q^{\mathcal{A}} d$. Put $e \triangleq (a \vee^{\mathfrak{A}} \neg^{\mathfrak{A}} a) \wedge^{\mathfrak{A}} (b \vee^{\mathfrak{A}} \neg^{\mathfrak{A}} b)$. Then, by (3.1), (3.7), (3.8), we have $(e \vee^{\mathfrak{A}} \neg^{\mathfrak{A}} e) = e$. Moreover, by (2.18), we get $e Q^{\mathcal{A}} c \wedge^{\mathfrak{A}} d$, so $c \wedge^{\mathfrak{A}} d \in Z^{\mathcal{A}}$. Finally, take any $a \in Z^{\mathcal{A}}$ and $c \in A$ such that $a \leqslant^{\mathfrak{A}} c$. Then, there is some $b \in A$ such that $(b \vee^{\mathfrak{A}} \neg^{\mathfrak{A}} b) Q^{\mathcal{A}} a$. Then, by (2.13) and (2.20), $(b \vee^{\mathfrak{A}} \neg^{\mathfrak{A}} b) Q^{\mathcal{A}} c$, so $c \in Z^{\mathcal{A}}$. Thus, $Z^{\mathcal{A}}$ is a filter of $\mathfrak{A}$, as required. ∎

Theorem 3.8. PCSQSA $= \mathbf{I}(\mathcal{S}_3 \times \mathcal{S}_2)$.

Proof. It is easy to see that $\mathcal{S}_3 \times \mathcal{S}_2$ is pseudo-complete, so $\mathbf{I}(\mathcal{S}_3 \times \mathcal{S}_2) \subseteq$ PCSQSA. For proving the converse, consider any $\mathcal{A} \in$ PCSQSA. Take any $\langle a, b \rangle \in A^2 \setminus Q^{\mathcal{A}}$, in which case, by (2.12), $\mathcal{A}$ is not singular. Then, by Theorem 3.4, there is some $h \in \mathrm{Hom}(\mathcal{A}, \mathcal{S}_3)$ such that $\langle a, b \rangle \notin h^{-1}[Q^{\mathcal{S}_3}]$. Moreover, as $\mathcal{A}$ is not singular, we have $\top^{\mathfrak{A}} \neq \bot^{\mathfrak{A}}$, and so, by (3.10), $\bot^{\mathfrak{A}} \notin Z^{\mathcal{A}}$. On the other hand, by (2.13), $Q^{\mathcal{A}}[Z^{\mathcal{A}}] \subseteq Z^{\mathcal{A}}$, so, $E^{\mathcal{A}}[Z^{\mathcal{A}}] \subseteq Z^{\mathcal{A}}$, in which case $Z^{\mathcal{A}} = (\nu^{\mathcal{A}})^{-1}[\nu^{\mathcal{A}}[Z^{\mathcal{A}}]]$. Therefore, by Lemma 3.7 and Remark 2.12, $\nu^{\mathcal{A}}[Z^{\mathcal{A}}]$ is a filter of $\mathcal{A}/E^{\mathcal{A}}$ not containing $\nu^{\mathcal{A}}(\bot^{\mathfrak{A}})$. Hence, by the Prime Ideal Theorem for distributive lattices, there is a prime filter Y of $\mathcal{A}/E^{\mathcal{A}}$ not containing $\nu^{\mathcal{A}}(\bot^{\mathfrak{A}})$ and including $Z^{\mathcal{A}}/E^{\mathcal{A}}$, in which case, by Remark 2.12, $X \triangleq (\nabla^{\mathcal{A}})^{-1}[Y]$ is a prime filter of $\mathcal{A}$ not containing $\bot^{\mathfrak{A}}$ and including $Z^{\mathcal{A}}$. Then, X is both proper and non-empty, in view of Lemma 3.7, so, by Lemma 3.3, $h^{\mathfrak{A}}_X \in \mathrm{Hom}(\mathfrak{A}, \mathfrak{S}_3)$, while $X = (h^{\mathfrak{A}}_X)^{-1}[3 \setminus 2]$, in which case $\varpi^X_{\mathcal{A}} = (h^{\mathfrak{A}}_X)^{-1}[Q^{\mathcal{S}_3}]$. On the other hand, as $\leqslant^{\mathcal{A}/E^{\mathcal{A}}}[Y] \subseteq Y$, by Remark 2.12, we have $Q^{\mathcal{A}}[X] \subseteq X$, so $Q^{\mathcal{A}} \subseteq \varpi^X_{\mathcal{A}}$. Thus, $h^{\mathfrak{A}}_X \in \mathrm{Hom}(\mathcal{A}, \mathcal{S}_3)$. And what is more, by (2.12), for every $c \in A$, $c \vee^{\mathfrak{A}} \neg^{\mathfrak{A}} c \in Z^{\mathcal{A}} \subseteq X$, in which case either $c \in X$ or $\neg^{\mathfrak{A}} c \in X$. Hence, $h^{\mathfrak{A}}_X \in \mathrm{Hom}(\mathcal{A}, \mathcal{S}_2)$. Finally, put $g \triangleq (h \times h^{\mathfrak{A}}_X)$. Then, $g \in \mathrm{Hom}(\mathcal{A}, \mathcal{S}_3 \times \mathcal{S}_2)$, while $\langle a, b \rangle \notin g^{-1}[Q^{\mathcal{S}_3 \times \mathcal{S}_2}]$. Thus, by Propositions 3.2 and 1.15, $\mathcal{A} \in \mathbf{I}(\mathcal{S}_3 \times \mathcal{S}_2)$, as required. ∎

Note that PSA $\subseteq$ SQSA, for $x_0 \in \Upsilon$.

Theorem 3.9. PSA $= \mathbf{I}(\langle \mathfrak{S}_3, \leqslant^{\mathfrak{S}_3} \rangle)$.

Proof. The inclusion from right to left is by Lemma 2.10. For proving the converse, consider any $\mathcal{A} \in$ PSA. Take any $\langle a, b \rangle \in A^2 \setminus Q^{\mathcal{A}}$. Then, by (2.20), $a \nleqslant^{\mathcal{A}} b$. Therefore, by the Prime Ideal Theorem for distributive lattices, there is a prime filter X of $\mathcal{A}$ such that $a \in X$, while $b \notin X$, in which case X is both non-empty and proper. Then, by Lemma 3.3, $h^{\mathfrak{A}}_X \in \mathrm{Hom}(\mathfrak{A}, \mathfrak{S}_3)$, in which case $h^{\mathfrak{A}}_X[\leqslant^{\mathfrak{A}}] \subseteq \leqslant^{\mathfrak{S}_3}$, so, by Lemma 2.10, $h^{\mathfrak{A}}_X \in \mathrm{Hom}(\mathcal{A}, \langle \mathfrak{S}_3, \leqslant^{\mathfrak{S}_3} \rangle)$. Moreover, $h^{\mathfrak{A}}_F a = 2$, while $h^{\mathfrak{A}}_F b \in 2$, so $h^{\mathfrak{A}}_F a \nleqslant^{\mathfrak{S}_3} h^{\mathfrak{A}}_F b$. Thus, by Propositions 3.2 and 1.15, $\mathcal{A} \in \mathbf{I}(\langle \mathfrak{S}_3, \leqslant^{\mathfrak{S}_3} \rangle)$, as required. ∎

Theorem 3.10. *Implicational classes of simple quasi-ordered Stone algebras form the five-element chain:*

(3.11) $$\text{OQSA} \subsetneq \text{SQBA} \subsetneq \text{PSA} \subsetneq \text{PCSQSA} \subsetneq \text{SQSA}.$$

Proof. First, the proper inclusion OQSA $\subsetneq$ SQBA is by Corollary 3.6. Next, given any $\mathcal{A} \in$ SQBA, since $E^{\mathcal{A}} \subseteq Q^{\mathcal{A}}$, by Proposition 2.7 and Lemma 3.1, we have $E^{\mathcal{A}} \in \mathrm{Con}(\mathcal{A})$, in which case $E^{\mathcal{A}} = \Delta_A$, for $\mathcal{A}$ is simple. Thus, SQBA $\subsetneq$

PSA, in view of Theorem 3.9, for $\mathfrak{S}_3$ is not a Boolean algebra. Further, by Lemma 2.10 and (3.2), PSA $\subseteq$ PCSQSA. On the other hand, (2.21) is not true in $\mathcal{S}_3 \times \mathcal{S}_2$ under the assignment $[x/\langle 1, 0\rangle, y/\langle 0, 0\rangle]$, so PSA $\neq$ PCSQSA, in view of Theorem 3.8. Finally, (3.10) is not true in $\mathcal{S}_3$ under the assignment $[x/1]$, so PCSQSA $\neq$ SQSA, in view of Theorem 3.4. Thus, the five subprevarieties of SQSA do form the chain (3.11). It only remains to argue that there is no more subprevariety of SQSA. For take any prevariety P $\subseteq$ SQSA. Consider the following five exhaustive cases:

1. P $\subseteq$ OQSA.
   Then, since OQSA $= \mathbf{I}(\varnothing) \subseteq$ P, we have P $=$ OQSA.

2. P $\subseteq$ SQBA but P $\not\subseteq$ OQSA.
   Take any non-singular $\mathcal{A} \in$ P. Then, $\bot^{\mathfrak{A}} \neq \top^{\mathfrak{A}}$. Therefore, by the inclusion SQBA $\subseteq$ PSA and Lemma 2.10, we have the embedding e of $\mathcal{S}_2$ into $\mathcal{A}$ defined by:

$$e0 \triangleq \bot^{\mathfrak{A}},$$
$$e2 \triangleq \top^{\mathfrak{A}}.$$

   Hence, $\mathcal{S}_2 \in$ P, so SQBA $\subseteq$ P, in view of Corollary 3.5. Thus, P $=$ SQBA.

3. P $\subseteq$ PSA but P $\not\subseteq$ SQBA.
   Take any $\mathcal{A} \in$ P not satisfying (3.6). Then, there is some $a \in A$ such that $b \triangleq a \vee^{\mathfrak{A}} \neg^{\mathfrak{A}} a \neq \top^{\mathfrak{A}}$, in which case $\mathcal{A}$ is not singular, while, by (3.1) and (3.7), we get $\neg^{\mathfrak{A}} b = \bot^{\mathfrak{A}}$, and so $b \neq \bot^{\mathcal{A}}$, in view of (3.2) and the non-singularity of $\mathcal{A}$. In this way, by Lemma 2.10, we have the embedding e of $\langle \mathfrak{S}_3, \leqslant^{\mathfrak{S}_3}\rangle$ into $\mathcal{A}$ defined by:

$$e0 \triangleq \bot^{\mathfrak{A}},$$
$$e1 \triangleq b,$$
$$e2 \triangleq \top^{\mathfrak{A}}.$$

   Hence, $\langle \mathfrak{S}_3, \leqslant^{\mathfrak{S}_3}\rangle \in$ P, so PSA $\subseteq$ P, in view of Theorem 3.9. Thus, P $=$ PSA.

4. P $\subseteq$ PCSQSA but P $\not\subseteq$ PSA.
   Take any $\mathcal{A} \in$ P not satisfying (2.21). Then, there is some $\langle a, b\rangle \in E^{\mathcal{A}}$ such that $a \neq b$, in which case $\mathcal{A}$ is not singular. Moreover, by (1.4), $\langle \neg^{\mathfrak{A}} a, \neg^{\mathfrak{A}} b\rangle \notin E^{\mathcal{A}}$, in which case, taking the symmetry between

a and b into account, without loss of generality, one can assume that $\langle\neg^{\mathfrak{A}}a, \neg^{\mathfrak{A}}b\rangle \notin Q^{\mathcal{A}}$, and so, by (2.20), $\neg^{\mathfrak{A}}a \nleqslant^{\mathfrak{A}} \neg^{\mathfrak{A}}b$. Then, once $\neg^{\mathfrak{A}}$ is a pseudocomplement operation, we get $c \triangleq \neg^{\mathfrak{A}}a \wedge^{\mathfrak{A}} b \neq \bot^{\mathfrak{A}}$. Moreover, by Lemma 2.8 and (3.1), we have $cE^{\mathcal{A}}(\neg^{\mathfrak{A}}a \wedge^{\mathfrak{A}} a) = \bot^{\mathfrak{A}}$, in which case, by Lemma 2.8, we also get $(c \vee^{\mathfrak{A}} \neg^{\mathfrak{A}}c)E^{\mathcal{A}}(\bot^{\mathfrak{A}} \vee^{\mathfrak{A}} \neg^{\mathfrak{A}}c) = \neg^{\mathfrak{A}}c$. On the other hand, as $\mathcal{A}$ is not singular, by (2.19) and (3.10), we then get $\langle\neg^{\mathfrak{A}}c, \bot^{\mathfrak{A}}\rangle \notin Q^{\mathcal{A}}$. Hence, by (2.13) and (2.20), we also conclude that $\langle\top^{\mathfrak{A}}, \bot^{\mathfrak{A}}\rangle \notin Q^{\mathcal{A}}$. Furthermore, $\langle\top^{\mathfrak{A}}, \neg^{\mathfrak{A}}c\rangle \notin Q^{\mathcal{A}}$, for, otherwise, by (3.2) and (2.20), we would have $\neg^{\mathfrak{A}}\bot^{\mathfrak{A}} = \top^{\mathfrak{A}}E^{\mathcal{A}}\neg^{\mathfrak{A}}c$, in which case, by (1.4), we would get $c = \bot^{\mathfrak{A}}$. In that case, we also have $\langle\neg^{\mathfrak{A}}\neg^{\mathfrak{A}}c, \neg^{\mathfrak{A}}c\rangle \notin Q^{\mathfrak{A}}$, for, otherwise, by (2.12), (2.19) and (3.5), we would have $\top^{\mathfrak{A}} = (\neg^{\mathfrak{A}}c \vee^{\mathfrak{A}} \neg^{\mathfrak{A}}\neg^{\mathfrak{A}}c)Q^{\mathfrak{A}}(\neg^{\mathfrak{A}}c \vee^{\mathfrak{A}} \neg^{\mathfrak{A}}c) = \neg^{\mathfrak{A}}c$. Hence, by (2.13) and (2.20), we also get $\langle\neg^{\mathfrak{A}}\neg^{\mathfrak{A}}c, \bot^{\mathfrak{A}}\rangle \notin Q^{\mathfrak{A}}$. Next, $\langle\top^{\mathfrak{A}}, \neg^{\mathfrak{A}}\neg^{\mathfrak{A}}c\rangle \notin Q^{\mathcal{A}}$, for, otherwise, by (3.2) and (2.20), we would have $\neg^{\mathfrak{A}}\bot^{\mathfrak{A}} = \top^{\mathfrak{A}}E^{\mathcal{A}}\neg^{\mathfrak{A}}\neg^{\mathfrak{A}}c$, in which case, by (3.2), (3.3) and (3.9), we would get $\neg^{\mathfrak{A}}c = \neg^{\mathfrak{A}}\neg^{\mathfrak{A}}\neg^{\mathfrak{A}}cQ^{\mathcal{A}}\neg^{\mathfrak{A}}\neg^{\mathfrak{A}}\bot^{\mathfrak{A}} = \bot^{\mathfrak{A}}$. Then, $\langle\neg^{\mathfrak{A}}c, \neg^{\mathfrak{A}}\neg^{\mathfrak{A}}c\rangle \notin Q^{\mathfrak{A}}$, for, otherwise, by (2.12), (2.19) and (3.5), we would have $\top^{\mathfrak{A}} = (\neg^{\mathfrak{A}}c \vee^{\mathfrak{A}} \neg^{\mathfrak{A}}\neg^{\mathfrak{A}}c)Q^{\mathfrak{A}}(\neg^{\mathfrak{A}}\neg^{\mathfrak{A}}c \vee^{\mathfrak{A}} \neg^{\mathfrak{A}}\neg^{\mathfrak{A}}c) = \neg^{\mathfrak{A}}\neg^{\mathfrak{A}}c$. In this way, as $c \leqslant^{\mathfrak{A}} \neg^{\mathfrak{A}}\neg^{\mathfrak{A}}c$, by (2.20), we have the strict homomorphism h from $\mathcal{S}_3 \times \mathcal{S}_2$ to $\mathcal{A}$ defined by:

$$
\begin{aligned}
h\langle 0,0\rangle &\triangleq \bot^{\mathfrak{A}},\\
h\langle 2,2\rangle &\triangleq \top^{\mathfrak{A}},\\
h\langle 0,2\rangle &\triangleq \neg^{\mathfrak{A}}c,\\
h\langle 2,0\rangle &\triangleq \neg^{\mathfrak{A}}\neg^{\mathfrak{A}}c,\\
h\langle 1,0\rangle &\triangleq c,\\
h\langle 1,2\rangle &\triangleq c \vee^{\mathfrak{A}} \neg^{\mathfrak{A}}c.
\end{aligned}
$$

Then, $\ker h \in \mathrm{Con}(\mathcal{S}_3 \times \mathcal{S}_2)$, so h is an embedding, for $\mathcal{S}_3 \times \mathcal{S}_2$ is simple. Hence, $\mathcal{S}_3 \times \mathcal{S}_2 \in \mathsf{P}$. Thus, by Theorem 3.8, $\mathsf{P} = \mathsf{PCSQSA}$.

5. $\mathsf{P} \nsubseteq \mathsf{PCSQSA}$.
Take any $\mathcal{A} \in \mathsf{P}$ not satisfying (3.10). Then, it is not singular, while there is some $a \in A$ such that $(a \vee^{\mathfrak{A}} \neg^{\mathfrak{A}}a)Q^{\mathcal{A}}\bot^{\mathfrak{A}}$, in which case, by (2.20), we get $(a \vee^{\mathfrak{A}} \neg^{\mathfrak{A}}a)E^{\mathcal{A}}\bot^{\mathfrak{A}}$. Moreover, since $\mathcal{A}$ is not singular, by (3.2), (3.3), (2.20) and (1.4), we also conclude that $\langle\top^{\mathfrak{A}}, \bot^{\mathfrak{A}}\rangle \notin Q^{\mathcal{A}}$. In this way, by (2.20), we have the strict homomorphism h from $\mathcal{S}_3$ to $\mathcal{A}$ defined by:

$$
\begin{aligned}
h0 &\triangleq \bot^{\mathfrak{A}},\\
h2 &\triangleq \top^{\mathfrak{A}},\\
h1 &\triangleq a \vee^{\mathfrak{A}} \neg^{\mathfrak{A}}a.
\end{aligned}
$$

Then, $\ker h \in \mathrm{Con}(\mathcal{S}_3)$, so h is an embedding, for $\mathcal{S}_3$ is simple. Hence, $\mathcal{S}_3 \in \mathsf{P}$. Thus, by Theorem 3.4, $\mathsf{P} = \mathsf{SQSA}$.

This completes the argument. ■

In view of Corollary 3.6, Theorem 3.10 implies:

Corollary 3.11. *Both* PSA *and* SQBA *are structurally complete. Moreover,*

$$\mathbf{V}(\mathsf{PSA}) = \mathbf{V}(\mathsf{SQSA}).$$

Thus, we have found all (in particular, axiomatic) extensions of the sequent calculus associated with $\mathcal{M}_3$ according to [31] (it had been proposed earlier in [20]; cf. [24], [35]) together with their finite relative axiomatizations. In this connection, we should like to highlight the following three points.

First, the extension, corresponding to PCSQSA, is relatively axiomatized by the single rule:

$$\frac{x \vdash; \neg x \vdash}{\vdash}.$$

Second, the extension, corresponding to PSA, is relatively axiomatized by the single rule:

$$\frac{\Gamma, x \vdash}{\Gamma \vdash \neg x}.$$

It has been introduced and studied in [24] and [35]. It is algebraizable with equivalent algebraic semantics SA (cf. Remark 2.11). Moreover, it is structurally complete (cf. [24]), so it is constituted exactly by all rules admissible in the initial sequent calculus. According to [30], it has the Cut-elimination property.

Third, the sequent calculus involved has the only proper consistent axiomatic extension that corresponds to $\mathsf{SQBA} \subseteq \mathsf{PSA}$ and is equivalent to BA (cf. Remark 2.11) and is relatively axiomatized by the single axiom:

$$\vdash x, \neg x.$$

It is deductively equivalent to Gentzen's calculus LK [8] for the classical logic and is structurally complete (cf. [24]).

3.2 Belnap's four-valued logic

Here, put $F \triangleq \{\wedge, \vee, \sim\}$ [optionally, $F = \widehat{F} \triangleq \{\wedge, \vee, \sim, \bot, \top\}$], where $\sim$ is unary (weak or constructive negation).

Recall that a *De Morgan lattice [algebra]* (cf. [2], [17], [27]) is any F-algebra, whose $(F \setminus \{\sim\})$-reduct is a [bounded] distributive lattice (cf. [2]) and the following identities are true in it:

$$\sim\sim x \approx x, \tag{3.12}$$

$$\sim(x \vee y) \approx \sim x \wedge \sim y, \tag{3.13}$$

$$\sim(x \wedge y) \approx \sim x \vee \sim y, \tag{3.14}$$

the variety of of all them being denoted by DML.[3] The following identities are true in De Morgan algebras:

$$\sim\bot \approx \top, \tag{3.15}$$

$$\sim\top \approx \bot, \tag{3.16}$$

A *Kleene lattice [algebra]* is any De Morgan lattice [algebra], satisfying the identity (cf. [27]):

$$(x \wedge \sim x) \lessapprox (y \vee \sim y). \tag{3.17}$$

The variety of all Kleene lattices [algebras] is denoted by KL. A *Boolean lattice [algebra]* is any De Morgan lattice [algebra] satisfying the identity (cf. [27]):

$$x \lessapprox (y \vee \sim y). \tag{3.18}$$

The variety of all Boolean lattices [algebras] is denoted by BL.

Remark 3.12. As it is well known, BL and BA are functionally rationally equivalent with respect to ι_F and the purely functional $(\widehat{F}, F)$-translation θ being the extension of ι_F defined by $\theta(\bot) \triangleq (x_0 \wedge \sim x_0)$ and $\theta(\top) \triangleq (x_0 \vee \sim x_0)$. ∎

A *quasi-ordered De Morgan (Kleene, Boolean) lattice [algebra]* is any quasi-ordered F-algebra, whose underlying algebra is a De Morgan (resp., Kleene, Boolean) lattice [algebra].

Set $\Upsilon \triangleq \{x_0, \sim x_0\}$.

Proposition 3.13. *Let $\mathcal{A}$ be a quasi-ordered De Morgan lattice [algebra]. Then, $\varepsilon^{\mathcal{A}} = \Im(\mathcal{A})$.*

Proof. By Lemma 2.8, (3.13) and (3.14), we have $\varepsilon^{\mathcal{A}} \in \mathrm{Con}(\mathcal{A})$. Moreover, by (3.12), we get $\varepsilon^{\mathcal{A}} \in \mathrm{Con}(\mathfrak{A})$. Then, Lemma 2.9 completes the argument. ∎

[3] We use here uniform notations, the meaning of which depends upon context.

Thus, by Remark 2.5 and Proposition 3.13, we see that a quasi-ordered De Morgan lattice [algebra] is simple iff it satisfies the quasi-identity (1.4). Hence, the class SQDML (SQKL , SQBL) of all simple quasi-ordered De Morgan (resp. Kleene, Boolean) lattices [algebras] is a quasivariety.

Put $\mathcal{DM}_4 \triangleq \langle \mathfrak{DM}_4, \varpi_{DM_4}^{(\pi_0)^{-1}[\{1\}]} \rangle$, where $\mathfrak{DM}_4 \upharpoonright (F \setminus \{\sim\}) \triangleq (\mathfrak{D}_2)^2$, whereas $\mathfrak{D}_2$ is the two-element [bounded] distributive lattice with carrier 2 and natural partial ordering given by $0 < 1$, while $\sim^{\mathfrak{DM}_4}\langle a, b\rangle \triangleq \langle 1 - b, 1 - a\rangle$, for all $a, b \in 2$. Moreover, the logic of the F-matrix $\mathcal{M}_4 \triangleq \langle \mathfrak{DM}_4, (\pi_0)^{-1}[\{1\}]\rangle$ with equality determinant Υ (cf. [31]) is Belnap's four-valued logic [1], [3], (cf. [22], [25]). Notice that $\mathcal{DM}_4 = \mathcal{M}_4{}^{[\{1\},\{1\}]}$. We will use the following standard abbreviations:

$$\begin{aligned} \mathsf{t} &:= \langle 1, 1\rangle, \\ \mathsf{f} &:= \langle 0, 0\rangle, \\ \mathsf{b} &:= \langle 1, 0\rangle, \\ \mathsf{n} &:= \langle 0, 1\rangle. \end{aligned}$$

Lemma 3.14. *Let $\mathfrak{A}$ be a De Morgan lattice [algebra] and X a [proper non-empty] prime filter of $\mathfrak{A}$. Then, we have the $h_X^{\mathfrak{A}} \in \mathrm{Hom}(\mathfrak{A}, \mathfrak{DM}_4)$ defined by $h_X^{\mathfrak{A}} \triangleq \chi_A^X \times (\chi_A^{A \setminus X} \circ \sim^{\mathfrak{A}})$.*

Proof. [As X is proper (non-empty), we have $\bot^{\mathfrak{A}} \notin X$ (resp., $\top^{\mathfrak{A}} \in X$).] Then, the identities (3.14), (3.13), (3.12)[, (3.15), (3.16)] and the fact that $\{1\}$ is a [proper non-empty] prime filter of $\mathfrak{D}_2$ complete the argument. ∎

Theorem 3.15. $\mathsf{SQDML} = \mathbf{I}(\mathcal{DM}_4)$.

Proof. Since $\mathfrak{DM}_4$ is a De Morgan lattice [algebra], while $(\pi_0)^{-1}[\{1\}]$ is a prime filter of $\mathcal{DM}_4$, $\mathcal{DM}_4$ is a quasi-ordered De Morgan lattice [algebra]. Moreover, for all $a, b \in DM_4$, $a = b$ iff both $\pi_0 a = \pi_0 b$ and $\pi_0 \sim^{\mathfrak{DM}_4} a = \pi_0 \sim^{\mathfrak{DM}_4} b$, while, for all $c, d \in 2$, $c = d$ iff $c = 1 \Leftrightarrow d = 1$. Therefore, (1.4) is true in $\mathcal{DM}_4$, so $\mathcal{DM}_4 \in \mathsf{SQDML}$, in which case $\mathbf{I}(\mathcal{DM}_4) \subseteq \mathsf{SQDML}$. For proving the converse, consider any $\mathcal{A} \in \mathsf{SQDML}$. Take an arbitrary $\langle a, b\rangle \in A^2 \setminus Q^{\mathcal{A}}$. Then, by Remark 2.12, $\nu^{\mathcal{A}}(a) \not\leqslant^{\mathcal{A}/E^{\mathcal{A}}} \nu^{\mathcal{A}}(b)$. Hence, by the Prime Ideal Theorem for distributive lattices, there is a prime filter Y of $\mathcal{A}/E^{\mathcal{A}}$ such that $\nu^{\mathcal{A}} a \in Y$, while $\nu^{\mathcal{A}}(b) \notin Y$, in which case, by Remark 2.12, $X \triangleq (\nu^{\mathcal{A}})^{-1}[Y]$ is a prime filter of $\mathcal{A}$ such that $a \in X$, while $b \notin X$, in which case [X is both proper and non-empty, whereas] $\langle a, b\rangle \notin \varpi_A^X$. Moreover, since $\leqslant^{\mathcal{A}/E^{\mathcal{A}}}[Y] \subseteq Y$, by Remark 2.12, we have $Q^{\mathcal{A}}[X] \subseteq X$, and so $Q^{\mathcal{A}} \subseteq \varpi_A^X$. Then, by Lemma 3.14, $h_X^{\mathfrak{A}} \in \mathrm{Hom}(\mathfrak{A}, \mathfrak{DM}_4)$, while $X = (h_X^{\mathfrak{A}})^{-1}[(\pi_0)^{-1}[\{1\}]]$, so $\varpi^X_A = (h^{\mathfrak{A}}_X)^{-1}[Q^{\mathcal{DM}_4}]$, in which case $\langle a, b\rangle \notin (h^{\mathfrak{A}}_X)^{-1}[Q^{\mathcal{DM}_4}]$, while $Q^{\mathcal{A}} \subseteq (h^{\mathfrak{A}}_X)^{-1}[Q^{\mathcal{DM}_4}]$, and so $h^{\mathfrak{A}}_X \in \mathrm{Hom}(\mathcal{A}, \mathcal{DM}_4)$. Thus, by Propositions 3.13 and 1.15, $\mathcal{A} \in \mathbf{I}(\mathcal{DM}_4)$, as required. ∎

Thus, in view of (1.2) and Theorem 3.15, for analyzing relative subvarieties of SQDML it suffices to analyze those of $\mathbf{S}(\mathcal{DM}_4)$.

The quasivariety of all one-element quasi-ordered De Morgan lattices [algebras] is denoted by OQDML. They are all trivial (cf. Remark 2.6). Clearly, $\mathsf{OQDML} = \mathbf{I}(\varnothing) \subseteq \mathsf{SQDML}$.

First of all, notice that any $B \subseteq DM_4$ forms a non-trivial substructure of $\mathcal{DM}_4$ iff

$$B \in \{DM_4, \{\mathsf{f}, \mathsf{t}\}, \{\mathsf{f}, \mathsf{t}, \mathsf{b}\}, \{\mathsf{f}, \mathsf{t}, \mathsf{n}\}\}.$$

Given any $a \in \{\mathsf{b}, \mathsf{n}\}$, put $\mathcal{DM}_a \triangleq \mathcal{DM}_4 \upharpoonright \{\mathsf{f}, \mathsf{t}, a\}$.

Note that any $\mathcal{B} \in \mathbf{S}(\mathcal{DM}_4)$ satisfies (3.17) iff either $B \subseteq \{\mathsf{f}, \mathsf{t}, \mathsf{b}\}$ or $B \subseteq \{\mathsf{f}, \mathsf{t}, \mathsf{n}\}$. Hence, by (1.2) and Theorem 3.15, we have:

Corollary 3.16. $\mathsf{SQKL} = \mathbf{I}(\{\mathcal{DM}_{\mathsf{b}}, \mathcal{DM}_{\mathsf{n}}\})$.

Note that any $\mathcal{B} \in \mathbf{S}(\mathcal{DM}_4)$ satisfies (3.18) iff $B \subseteq \{\mathsf{f}, \mathsf{t}\}$. Put $\mathcal{DM}_2 \triangleq \mathcal{DM}_4 \upharpoonright \{\mathsf{f}, \mathsf{t}\}$. Hence, by (1.2) and Theorem 3.15, we have:

Corollary 3.17. $\mathsf{SQBL} = \mathbf{I}(\mathcal{DM}_2)$.

Remark 3.18. Notice that $\mathcal{DM}_2$ is partially-ordered, and so is any member of SQBL, in view of Corollary 3.17. Hence, by Remark 2.11, SQBL is relationally rationally equivalent to BL with respect to translations involved therein. ∎

A quasi-ordered De morgan lattice [algebra] is said to be *non-paraconsistent*, if it satisfies the following identity:

$$(x \wedge {\sim}x)Qy. \tag{3.19}$$

The quasivariety of all simple non-paraconsistent quasi-ordered De Morgan lattices [algebras] is denoted by NPSQDML. Note that any $\mathcal{B} \in \mathbf{S}(\mathcal{DM}_4)$ satisfies (3.19) iff $B \subseteq DM_{\mathsf{n}}$. Hence, by (1.2) and Theorem 3.15, we have:

Corollary 3.19. $\mathsf{NPSQDML} = \mathbf{I}(\mathcal{DM}_{\mathsf{n}})$.

Remark 3.20. It is easy ckecking that $\mathcal{DM}_{\mathsf{n}}$ is relationally rationally equivalent to $\mathfrak{DM}_{\mathsf{n}}$ with respect to the purely relational $(L, F + \approx)$-translation ρ_{n}, defined by $\rho_{\mathsf{n}}(Q) \triangleq \{x_0 \lessapprox (x_1 \vee (x_0 \wedge {\sim}x_0))\}$ and ι_F. Hence, as $\mathsf{KL} = \mathbf{I}(\mathfrak{DM}_4 \upharpoonright \{\mathsf{f}, \mathsf{t}, \mathsf{n}\})$ (cf. [12], [27]), by Theorem 1.4(iv)⇒(iii) and Corollary 3.19, NPSQDML and KL are relationally rationally equivalent with respect to ρ_{n} and ι_F. ∎

A quasi-ordered De Morgan lattice [algebra] is said to be *non-paracomplete*, if it satisfies the following identity:

$$yQ(x \vee \sim x). \tag{3.20}$$

The quasivariety of all simple non-paracomplete quasi-ordered De Morgan lattices [algebras] is denoted by $\mathsf{NCSQDML}$. Note that any $\mathcal{B} \in \mathbf{S}(\mathcal{DM}_4)$ satisfies (3.20) iff $B \subseteq DM_{\mathsf{b}}$. Hence, by (1.2) and Theorem 3.15, we have:

Corollary 3.21. $\mathsf{NCSQDML} = \mathbf{I}(\mathcal{DM}_{\mathsf{b}})$.

Remark 3.22. It is easy ckecking that $\mathcal{DM}_{\mathsf{b}}$ is relationally rationally equivalent to $\mathfrak{DM}_{\mathsf{b}}$ with respect to the purely relational $(L, F + \approx)$-translation ρ_{b}, defined by $\rho_{\mathsf{b}}(Q) \triangleq \{x_0 \wedge (x_1 \vee \sim x_1) \lessapprox x_1\}$ and ι_F. Hence, as $\mathsf{KL} = \mathbf{I}(\mathfrak{DM}_4 \upharpoonright \{\mathsf{f}, \mathsf{t}, \mathsf{b}\})$ (cf. [12], [27]), by Theorem 1.4(iv)⇒(iii) and Corollary 3.21, $\mathsf{NCSQDML}$ and KL are relationally rationally equivalent with respect to ρ_{b} and ι_F. ∎

In view of Corollaries 3.16, 3.19, 3.21, from now on, non-paraconsistent (non-paracomplete) simple quasi-ordered De Morgan lattices [algebras] are referred to as *non-paraconsistent (non-paracomplete) simple quasi-ordered Kleene lattices [algebras]*, the quasivariety of all them being denoted by NPSQKL (resp., by NCSQKL).

As any relative subvariety of any class is closed under $\mathbf{S}$, by (1.2) and Theorem 3.15, we eventually get:

Corollary 3.23. *Relative subvarieties of* SQDML *form the six-element non-chain distributive lattice depicted at Figure 3.1.*

3.2.1 Partially-ordered De Morgan lattices and algebras

The quasivariety of all partially-ordered De Morgan (Kleene) lattices [algebras] is denoted by PDML (resp., by PKL). Recall that they are simple. Moreover, by Lemma 2.10 and Remark 3.18, we also have:

$$\mathsf{NPSQKL} \cap \mathsf{PDML} = \mathsf{SQBL} = \mathsf{NCSQKL} \cap \mathsf{PDML}. \tag{3.21}$$

Proposition 3.24. PDML *is the subquasivariety of* SQDML *relatively axiomatized by the single quasi-identity:*

$$(xQy) \to (\sim yQ \sim x). \tag{3.22}$$

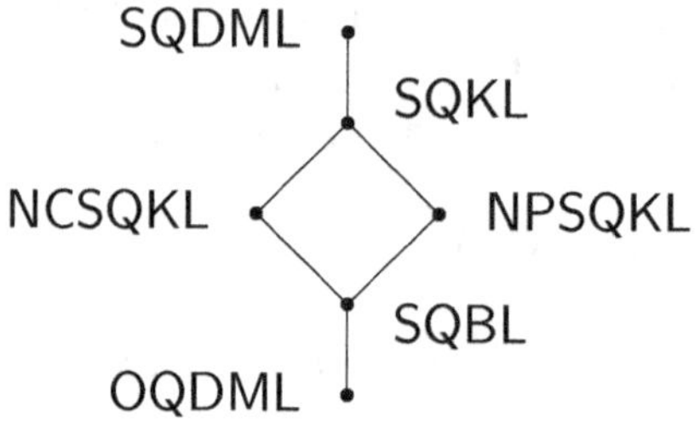

Figure 3.1: The lattice of relative subvarieties of SQDML.

Proof. The fact that (3.22) is true in PDML is by Lemma 2.10. Conversely, take any $\mathcal{A} \in$ SQDML satisfying (3.22). Then, in view of (1.4), (2.21) is true in $\mathcal{A}$, as required. ∎

Proposition 3.25. $\mathsf{PDML} = \mathbf{I}(\langle \mathfrak{DM}_4, \leqslant^{\mathfrak{DM}_4} \rangle)$.

Proof. As $\mathfrak{DM}_4$ is a De Morgan lattice, Lemma 2.10 yields the inclusion from right to left. Conversely, take any $\mathcal{A} \in$ PDML. Consider an arbitrary $\langle a, b \rangle \notin Q^{\mathcal{A}}$. Then, by Lemma 2.10, $a \not\leqslant^{\mathfrak{A}} b$. Then, by the Prime Ideal Theorem, there is a prime filter X of $\mathcal{A}$ such that $a \in X$, while $b \notin X$[, in which case X is both non-empty and proper]. Then, by Lemma 3.14, $h^{\mathcal{A}}_X \in \mathrm{Hom}(\mathfrak{A}, \mathfrak{DM}_4)$, in which case $h^{\mathcal{A}}_X[\leqslant^{\mathfrak{A}}] \subseteq \leqslant^{\mathfrak{DM}_4}$, so, by Lemma 2.10, $h^{\mathcal{A}}_X \in \mathrm{Hom}(\mathcal{A}, \langle \mathfrak{DM}_4, \leqslant^{\mathfrak{DM}_4} \rangle)$. Moreover, $X = (h^{\mathfrak{A}}_X)^{-1}[\{\mathsf{t}, \mathsf{b}\}]$, so $h^{\mathfrak{A}}_X(a) \not\leqslant^{\mathfrak{DM}_4} h^{\mathfrak{A}}_X(b)$. Then, by Propositions 3.13 and 1.15, $\mathcal{A} \in \mathbf{I}(\langle \mathfrak{DM}_4, \leqslant^{\mathfrak{DM}_4} \rangle)$, as required. ∎

Notice that $\mathfrak{DM}_{\mathsf{b}}$ and $\mathfrak{DM}_{\mathsf{n}}$ are isomorphic. Hence, by (1.2) and Theorem 3.25, we have:

Corollary 3.26. $\mathsf{PKL} = \mathbf{I}(\langle \mathfrak{DM}_{\mathsf{b}}, \leqslant^{\mathfrak{DM}_{\mathsf{b}}} \rangle)$.

3.2.2 Expansions of non-idempotent De Morgan lattices

Let $L' = \langle F', R' \rangle$ be a first-order signature such that $F \subseteq F'$. An L'-structure is said to be *non-idempotent*, provided it satisfies the following quasi-identity:

(3.23) $$(\sim x \approx x) \to (x \approx y).$$

The quasivariety of all non-idempotent L'-structures is denoted by $\mathsf{NI}_{L'}$.

Let $\mathcal{B}$ be an L'-structure such that $\mathcal{B} \upharpoonright F = \mathfrak{DM}_4$, while $\{\mathsf{f}, \mathsf{t}\}$ forms a substructure of $\mathcal{B}$, whereas any singular substructure of $\mathcal{B}$ is trivial.

The following generic auxiliary statement properly inherits Lemma 4.3 of [27] together with its proof, though does not follow from the latter immediately, simply because, first, an expansion of a De Morgan lattice [algebra] need not be generated by the same subset as its De Morgan lattice [algebra] reduct is, while, second, homomorphisms between underlying algebras need not be homomorphisms between initial algebraic systems.

Lemma 3.27. *Let L' and $\mathcal{B}$ be as above, Then, for any non-trivial, finitely-generated and non-idempotent $\mathcal{A} \in \mathbf{I}(\mathcal{B})$, it holds that* $\mathrm{Hom}(\mathcal{A}, \mathcal{B} \upharpoonright \{\mathsf{f}, \mathsf{t}\}) \neq \varnothing$.

Proof. By contradiction. For take any non-trivial, finitely-generated and non-idempotent $\mathcal{A} \in \mathbf{I}(\mathcal{B})$ such that $\mathrm{Hom}(\mathcal{A}, \mathcal{B} \upharpoonright \{f, \mathsf{t}\}) = \varnothing$. Assume $\mathcal{A}$ is generated by some finite non-empty[4] $\{a_1, \ldots, a_n\} \subseteq A$, where $n \geqslant 1$. By induction on $1 \leqslant i \leqslant n$, define b_i as follows:

$$b_1 \triangleq a_1 \vee^{\mathfrak{A}} \sim^{\mathfrak{A}} a_1,$$
$$b_{j+1} \triangleq (a_{j+1} \vee^{\mathfrak{A}} \sim^{\mathfrak{A}} a_{j+1} \vee^{\mathfrak{A}} \sim^{\mathfrak{A}} b_j) \wedge^{\mathfrak{A}} b_j,$$

for all $1 \leqslant j \leqslant n-1$. Consider any $h \in \mathrm{Hom}(\mathcal{A}, \mathcal{B})$. By induction on $1 \leqslant i \leqslant n$, one can check that $hb_i \in \{\mathsf{n}, \mathsf{b}, \mathsf{t}\}$. As $\mathrm{Hom}(\mathcal{A}, \mathcal{B} \upharpoonright \{\mathsf{f}, \mathsf{t}\}) = \varnothing$, there must be some $1 \leqslant i_h \leqslant n$ such that $ha_{i_h} \in \{\mathsf{n}, \mathsf{b}\}$. By induction on $i_h \leqslant j \leqslant n$, one can check that $hb_j \in \{\mathsf{n}, \mathsf{b}\}$ as well. In particular, $hb_n \in \{\mathsf{n}, \mathsf{b}\}$, in which case $h{\sim}^{\mathfrak{A}} b_n = hb_n$, for each $h \in \mathrm{Hom}(\mathcal{A}, \mathcal{B})$. Therefore, as $\mathcal{A} \in \mathbf{I}(\mathcal{B})$, ${\sim}^{\mathfrak{A}} b_n = b_n$. In view of (3.23), this implies that $\mathcal{A}$ is singular, in which case, by Proposition 1.2, it is trivial. This contradiction to the initial assumption completes the argument. ■

Theorem 3.28. *Let L' and $\mathcal{B}$ be as above, and $\mathsf{K} \subseteq \mathbf{I}(\mathcal{B})$. Suppose both K and each member of it are finite, while $\mathcal{B} \upharpoonright \{\mathsf{f}, \mathsf{t}\} \in \mathbf{I}(\mathsf{K})$. Then, $\mathbf{I}(\mathsf{K}) \cap \mathsf{NI}_{L'} = \mathbf{I}(\{\mathcal{C} \times (\mathcal{B} \upharpoonright \{\mathsf{f}, \mathsf{t}\}) | \mathcal{C} \in \mathsf{K}\})$.*

Proof. First of all, notice that, for any L'-structure $\mathcal{C}$, $\mathcal{C} \times (\mathcal{B} \upharpoonright \{\mathsf{f}, \mathsf{t}\})$ is non-idempotent. This yields the inclusion from right to left. Further, as both $\{\mathcal{C} \times (\mathcal{B} \upharpoonright \{\mathsf{f}, \mathsf{t}\}) | \mathcal{C} \in \mathsf{K}\}$ and each member of it are finite, the prevariety generated by it is a quasivariety, so it suffices to prove the inverse inclusion

[4] Recall that carriers of algebraic systems are always supposed to be non-empty.

for finitely-generated structures alone. For take any non-idempotent finitely-generated $\mathcal{A} \in \mathbf{I}(\mathsf{K})$. Consider any $r \in (R' + \approx)$ and an arbitrary $\overline{a} \in A^{\mu_{L'+\approx}(r)} \setminus r^{\mathfrak{A}+\approx}$, in which case $\mathcal{A}$ is non-trivial. Then, there are some $\mathcal{C} \in \mathsf{K}$ and $h \in \mathrm{Hom}(\mathcal{A}, \mathcal{C})$ such that $\overline{a} \notin h^{-1}[r^{\mathcal{C}+\approx}]$. Moreover, by Lemma 3.27, there is some $g \in \mathrm{Hom}(\mathcal{A}, \mathcal{B} \upharpoonright \{\mathsf{f}, \mathsf{t}\})$, in which case $(h \times g) \in \mathrm{Hom}(\mathcal{A}, \mathcal{C} \times (\mathcal{B} \upharpoonright \{\mathsf{f}, \mathsf{t}\}))$, while $\overline{a} \notin (h \times g)^{-1}[r^{(\mathcal{C} \times (\mathcal{B} \upharpoonright \{\mathsf{f},\mathsf{t}\}))+\approx}]$. Thus, $\mathcal{A} \in \mathbf{I}(\{\mathcal{C} \times (\mathcal{B} \upharpoonright \{\mathsf{f}, \mathsf{t}\}) | \mathcal{C} \in \mathsf{K}\})$, as required. ∎

As a particular case of Theorem 3.28 with $\mathsf{K} = \{\mathcal{B}\}$, we get:

Corollary 3.29. *Let L' and $\mathcal{B}$ be as above. Then,* $\mathbf{I}(\mathcal{B}) \cap \mathsf{NI}_{L'} = \mathbf{I}(\mathcal{B} \times (\mathcal{B} \upharpoonright \{\mathsf{f}, \mathsf{t}\}))$.

Recall that $\mathsf{DML} = \mathbf{I}(\mathfrak{D}\mathfrak{M}_4)$ (cf. [12], [27]), so the above generic results do deal with arbitrary expansions of non-idempotent De Morgan lattices.

The quasivariety of all non-idempotent simple quasi-ordered De Morgan (Kleene) lattices [algebras] is denoted by $\mathsf{NISQDML}$ (resp., by NISQKL). Corollary 3.29 with $L' = L$ together with Remark 2.6 and Theorem 3.15 immediately yield:

Proposition 3.30. $\mathsf{NISQDML} = \mathbf{I}(\mathcal{DM}_4 \times \mathcal{DM}_2)$.

Theorem 3.28 with $L' = L$ together with Remark 2.6 and Corollary 3.16 immediately yield:

Proposition 3.31. $\mathsf{NISQKL} = \mathbf{I}(\{\mathcal{DM}_{\mathsf{b}} \times \mathcal{DM}_2, \mathcal{DM}_{\mathsf{n}} \times \mathcal{DM}_2\})$.

The quasivariety of all non-idempotent non-paraconsistent (non-paracomplete) simple quasi-ordered Kleene lattices [algebras] is denoted by $\mathsf{NINPSQKL}$ (resp., by $\mathsf{NINCSQKL}$).

Theorem 3.28 with $L' = L$ together with Remark 2.6 and Corollaries 3.19 and 3.21 respectively imply:

Proposition 3.32. $\mathsf{NINPSQKL} = \mathbf{I}(\mathcal{DM}_{\mathsf{n}} \times \mathcal{DM}_2)$.

Proposition 3.33. $\mathsf{NINCSQKL} = \mathbf{I}(\mathcal{DM}_{\mathsf{b}} \times \mathcal{DM}_2)$.

The quasivariety of all non-idempotent partially-ordered De Morgan (KLeene) lattices [algebras] is denoted by NIPDML (resp., by NIPKL).

First, Corollary 3.29 with $L' = L$ together with Remarks 2.6, 3.18, Proposition 3.25 directly imply:

Corollary 3.34. $\mathsf{NIPDML} = \mathbf{I}(\langle \mathfrak{D}\mathfrak{M}_4, \leqslant^{\mathfrak{D}\mathfrak{M}_4} \rangle \times \mathcal{DM}_2)$.

Next, Theorem 3.28 with $L' = L$ together with Remarks 2.6, 3.18, Corollary 3.26 directly imply:

Corollary 3.35. $\mathsf{NIPKL} = \mathbf{I}(\langle \mathfrak{DM}_{\mathsf{b}}, \leqslant^{\mathfrak{DM}_{\mathsf{b}}} \rangle \times \mathcal{DM}_2)$.

We have provided above general tools of studying meets with $\mathbf{I}(\mathcal{B}) \cap \mathsf{NI}_{L'}$. It appears that there are equally generic tools of studying joins with $\mathbf{I}(\mathcal{B}) \cap \mathsf{NI}_{L'}$.

Let $\mathbb{R}$ be a universal Horn $(L' + \approx)$-theory. Given any $R = (\Gamma \to \Phi) \in \mathbb{R}$, choose any $v_R \in V^\infty \setminus \mathrm{Var}(R)$ and put $\mathrm{NI}(R) \triangleq ((\{{\sim} v_R \approx v_R\} \cup \Gamma) \to \Phi)$. After all, set $\mathrm{NI}(\mathbb{R}) \triangleq \{\mathrm{NI}(R) | R \in \mathbb{R}\}$.

Lemma 3.36. *Let L' and $\mathcal{B}$ be as above, while P the subprevariety of $\mathbf{I}(\mathcal{B})$ relatively axiomatized by a universal Horn $(L' + \approx)$-theory $\mathbb{R}$ and generated by a class of L'-structures K. Then, $(\mathbf{I}(\mathcal{B}) \cap \mathsf{NI}_{L'}) \cup \mathsf{P}$ is the subprevariety of $\mathbf{I}(\mathcal{B})$ relatively axiomatized by $\mathrm{NI}(\mathbb{R})$ and generated by $\mathsf{K} \cup \{\mathcal{B} \times (\mathcal{B} \upharpoonright \{\mathsf{f}, \mathsf{t}\})\}$.*

Proof. Consider any $R = (\Gamma \to \Phi) \in \mathbb{R}$. Then, $\mathrm{NI}(R)$ is a logical consequence of R, so the former is true in P, for the latter is. Hence, $\mathrm{NI}(\mathbb{R})$ is true in P. Further, as, by Proposition 1.2, any singular structure in $\mathbf{I}(\mathcal{B})$ is trivial, by (3.23), the quasi-identity $({\sim} v_R \approx v_R) \to \Phi$ is true in $\mathbf{I}(\mathcal{B}) \cap \mathsf{NI}_{L'}$, in which case the implication $\mathrm{NI}(R)$, being a logical consequence of the quasi-identity involved, is so. Thus, $\mathrm{NI}(\mathbb{R})$ is true in $\mathbf{I}(\mathcal{B}) \cap \mathsf{NI}_{L'}$. Conversely, take any $\mathcal{A} \in \mathbf{I}(\mathcal{B}) \cap \mathrm{Mod}_{\approx}(\mathrm{NI}(\mathbb{R}))$. Assume $\mathcal{A} \notin \mathsf{NI}_{L'}$. Then, there is some $a \in A$ such that ${\sim}^{\mathfrak{A}} a = a$. Consider any $R \in \mathbb{R}$ and any assignment $h : \mathrm{Var}(R) \to A$. Extend the latter to the $g : \mathrm{Var}(\mathrm{NI}(R)) \to A$ by setting $g(v_R) \triangleq a$. Then, $\mathcal{A} \models \mathrm{NI}(R)[g]$, in which case $\mathcal{A} \models R[h]$. Hence, $\mathcal{A} \in \mathsf{P}$. Therefore, $\mathcal{A} \in (\mathbf{I}(\mathcal{B}) \cap \mathsf{NI}_{L'}) \cup \mathsf{P}$. Then, Corollary 3.29 completes the argument. ∎

First, by Lemma 3.36, Remark 2.6, Proposition 3.30 and Corollary 3.16, we immediately have:

Corollary 3.37. (cf. Corollary 4.4 of [27]) $\mathsf{NISQDML} \cup \mathsf{SQKL}$ *is the subquasivariety of* SQDML *relatively axiomatized by the single quasi-identity:*

$$({\sim} z \approx z) \to ((x \wedge {\sim} x) \lessapprox (y \vee {\sim} y)). \tag{3.24}$$

Moreover, $\mathsf{NISQDML} \cup \mathsf{SQKL} = \mathbf{I}(\{\mathcal{DM}_4 \times \mathcal{DM}_2, \mathcal{DM}_{\mathsf{b}}, \mathcal{DM}_{\mathsf{n}}\})$.

As a consequence, we also get:

Corollary 3.38. (cf. Corollary 4.4 of [27]) $\mathsf{NIPDML} \cup \mathsf{PKL}$ *is the subquasivariety of* PDML *relatively axiomatized by the single quasi-identity* (3.24). *Moreover,* $\mathsf{NIPDML} \cup \mathsf{PKL} = \mathbf{I}(\{\langle \mathfrak{DM}_4, \leqslant^{\mathfrak{DM}_4} \rangle \times \mathcal{DM}_2, \langle \mathfrak{DM}_{\mathsf{b}}, \leqslant^{\mathfrak{DM}_{\mathsf{b}}} \rangle\})$.

Proof. Note that $\mathsf{PDML} \cap (\mathsf{NISQDML} \cup \mathsf{SQKL}) = \mathsf{NIPDML} \cup \mathsf{PKL}$. Then, Corollaries 3.37, 3.34 and 3.26 complete the argument ■

Next, by Lemma 3.36, Remark 2.6, Propositions 3.30 and 3.25, we immediately have:

Corollary 3.39. $\mathsf{NISQDML} \cup \mathsf{PDML}$ *is the subquasivariety of* SQDML *relatively axiomatized by the single quasi-identity:*

$$(\sim z \approx z)\&(xQy)\&(yQx) \to (x \approx y). \tag{3.25}$$

Moreover, $\mathsf{NISQDML} \cup \mathsf{PDML} = \mathbf{I}(\{\mathcal{DM}_4 \times \mathcal{DM}_2, \langle \mathfrak{DM}_4, \leqslant^{\mathfrak{DM}_4} \rangle)$.

Further, by Lemma 3.36, Remark 2.6, Proposition 3.30 and Corollary 3.19, we immediately get:

Corollary 3.40. $\mathsf{NISQDML} \cup \mathsf{NPSQKL}$ *is the subquasivariety of* SQDML *relatively axiomatized by the single quasi-identity:*

$$(\sim z \approx z) \to ((x \wedge \sim x)Qy). \tag{3.26}$$

Moreover, $\mathsf{NISQDML} \cup \mathsf{NPSQKL} = \mathbf{I}(\{\mathcal{DM}_4 \times \mathcal{DM}_2, \mathcal{DM}_\mathsf{n}\})$.

As a consequence, we also get:

Corollary 3.41. $\mathsf{NISQKL} \cup \mathsf{NPSQKL}$ *is the subquasivariety of* SQKL *relatively axiomatized by the single quasi-identity* (3.26). *Moreover,* $\mathsf{NISQKL} \cup \mathsf{NPSQKL} = \mathbf{I}(\{\mathcal{DM}_\mathsf{b} \times \mathcal{DM}_2, \mathcal{DM}_\mathsf{n}\})$.

Proof. First of all, we have $\mathsf{SQKL} \cap (\mathsf{NISQDML} \cup \mathsf{NPSQKL}) = \mathsf{NISQKL} \cup \mathsf{NPSQKL}$. Then, Corollaries 3.40, 3.19, 3.31 complete the argument, for $\mathcal{DM}_\mathsf{n} \times \mathcal{DM}_2 \in \mathbf{I}(\mathcal{DM}_\mathsf{n})$. ■

Finally, by Lemma 3.36, Remark 2.6, Proposition 3.30 and Corollary 3.21, we immediately get:

Corollary 3.42. $\mathsf{NISQDML} \cup \mathsf{NCSQKL}$ *is the subquasivariety of* SQDML *relatively axiomatized by the single quasi-identity:*

$$(\sim z \approx z) \to (yQ(x \vee \sim x)). \tag{3.27}$$

Moreover, $\mathsf{NISQDML} \cup \mathsf{NCSQKL} = \mathbf{I}(\{\mathcal{DM}_4 \times \mathcal{DM}_2, \mathcal{DM}_\mathsf{b}\})$.

As a consequence, we also get:

Corollary 3.43. $\mathsf{NISQKL} \cup \mathsf{NCSQKL}$ *is the subquasivariety of* SQKL *relatively axiomatized by the single quasi-identity* (3.27). *Moreover,* $\mathsf{NISQKL} \cup \mathsf{NCSQKL} = \mathbf{I}(\{\mathcal{DM}_{\mathsf{n}} \times \mathcal{DM}_2, \mathcal{DM}_{\mathsf{b}}\})$.

Proof. First of all, we have $\mathsf{SQKL} \cap (\mathsf{NISQDML} \cup \mathsf{NCSQKL}) = \mathsf{NISQKL} \cup \mathsf{NCSQKL}$. Then, Corollaries 3.42, 3.21, 3.31 complete the argument, for $\mathcal{DM}_{\mathsf{b}} \times \mathcal{DM}_2 \in \mathbf{I}(\mathcal{DM}_{\mathsf{b}})$. ∎

3.2.3 Regular simple quasi-ordered Kleene lattices and algebras

A simple quasi-ordered Kleene lattice [algebra] is said to be *regular*, if it satisfies the quasi-identity (cf. [27]):

$$((\sim x \lessapprox x)\&((x \wedge \sim y) \lessapprox (\sim x \vee y))) \to (\sim y \lessapprox y). \tag{3.28}$$

Proposition 3.44. *The following hold:*

(i) *any regular* $\mathcal{A} \in \mathsf{SQKL}$ *is non-idempotent;*

(ii) *any* $\mathcal{A} \in \mathsf{SQBL}$ *is regular.*

Proof. (i) Take any $a, b \in A$. Suppose $\sim^{\mathfrak{A}} a = a$. Then, applying (3.28) under the assignment $[x/a, y/(a \wedge^{\mathcal{A}} b \wedge^{\mathfrak{A}} \sim^{\mathcal{A}} b)]$, by (3.14) and (3.12), we have $(a \vee^{\mathcal{A}} b) \leqslant^{\mathfrak{A}} (a \vee^{\mathcal{A}} \sim^{\mathcal{A}} b \vee^{\mathcal{A}} b) \leqslant^{\mathcal{A}} (a \wedge^{\mathcal{A}} b \wedge^{\mathfrak{A}} \sim^{\mathcal{A}} b) \leqslant^{\mathcal{A}} (a \wedge^{\mathcal{A}} b)$, in which case we get both $a \leqslant^{\mathfrak{A}} b$ and $b \leqslant^{\mathfrak{A}} a$, and so $a = b$. Thus, (3.23) is true in $\mathcal{A}$, as required.

(ii) Take any $a, b \in A$. Suppose $\sim^{\mathcal{A}} a \leqslant^{\mathcal{A}} a$ and $a \wedge^{\mathcal{A}} \sim^{\mathfrak{A}} b \leqslant^{\mathcal{A}} \sim^{\mathcal{A}} a \vee^{\mathcal{A}} b$. Then, $\sim^{\mathcal{A}} a \vee^{\mathcal{A}} a = a$, so, by (3.18), we have $a \wedge^{\mathcal{A}} \sim^{\mathfrak{A}} b = \sim^{\mathfrak{A}} b$, in which case, by (3.14) and (3.12), we also get $\sim^{\mathcal{A}} a \vee^{\mathcal{A}} b = b$. Hence, $\sim^{\mathfrak{A}} b \leqslant^{\mathcal{A}} b$. Thus, (3.28) is true in $\mathcal{A}$, as required. ∎

The quasivariety of all ({non-paraconsistent,non-paracomplete}) regular (simple) partially-(quasi-)ordered Kleene lattices [algebras] is denoted by RPKL (RSQKL {resp., $\mathsf{RNPSQKL}$, $\mathsf{RNCSQKL}$}).

Given any $a \in \{\mathsf{n}, \mathsf{b}\}$, the set $4_a \triangleq \{\langle \mathsf{f}, \mathsf{f} \rangle, \langle a, \mathsf{f} \rangle, \langle a, \mathsf{t} \rangle, \langle \mathsf{t}, \mathsf{t} \rangle\}$ forms a subalgebra of $\mathfrak{DM}_a \times \mathfrak{DM}_2$, satisfying (3.28). An L-structure $\mathcal{A}$ is said to be *pre-regular*, whenever $\mathcal{A} = \mathcal{C} \times \mathcal{DM}_2$, where either $\mathcal{C} = \mathcal{DM}_a$ or $\mathcal{C} = \langle \mathfrak{DM}_a, \leqslant^{\mathfrak{DM}_a} \rangle$, whereas $a \in \{\mathsf{n}, \mathsf{b}\}$, in which case $\mathcal{A} \downarrow 4 \triangleq \mathcal{A} \upharpoonright 4_a$ is a regular simple quasi-ordered Kleene lattice. For any class K of pre-regular L-structures, set $\mathsf{K} \downarrow 4 \triangleq \{\mathcal{A} \downarrow 4 | \mathcal{A} \in \mathsf{K}\}$.

Lemma 3.45. *Let* $\mathsf{K} \subseteq \mathsf{SQKL}$ *be a class of pre-regular L-structures. Then,* $\mathsf{RSQKL} \cap \mathbf{I}(\mathsf{K}) = \mathbf{I}(\mathsf{K} \downarrow 4)$.

Proof. The inclusion from right to left is obvious. As both $\mathsf{K} \downarrow 4$ and each member of it are finite, for K has at most four structures, each being finite, $\mathbf{I}(\mathsf{K} \downarrow 4)$ is a quasivariety, so it suffices to prove the converse inclusion for finitely-generated structures alone. For take any finitely-generated regular $\mathcal{A} \in \mathbf{I}(\mathsf{K})$. Assume $\overline{a} \in A^n$, where $n \in \omega \setminus 0$, while $\mathcal{A}$ is generated by $\operatorname{img} \overline{a}$. Take any $\mathcal{B} \in \mathsf{K}$, in which case $\mathcal{B} = \mathcal{C} \times \mathcal{DM}_2$, while $\mathfrak{C} = \mathfrak{DM}_l$, where $l \in \{\mathsf{n}, \mathsf{b}\}$. Consider an arbitrary $h \in \mathrm{Hom}(\mathcal{A}, \mathcal{B})$. Put $I_h \triangleq \{i \in n | ha_i \leqslant \langle \mathsf{f}, \mathsf{t} \rangle\}$ and $J_h \triangleq \{j \in n | ha_j \geqslant \langle \mathsf{t}, \mathsf{f} \rangle\}$. (Note that $I_h \cap J_h = \varnothing$.) By contradiction, let us prove that there is some $g \in \mathrm{Hom}(\mathcal{A}, \mathcal{DM}_2)$ such that, for every $i \in I_h$, $ga_i = \mathsf{f}$, while, for every $j \in J_h$, $ga_j = \mathsf{t}$. For suppose that, for each $g \in \mathrm{Hom}(\mathcal{A}, \mathcal{DM}_2)$, there is either some $i \in I_h$ such that $ga_i = \mathsf{t}$ or some $j \in J_h$ such that $ga_j = \mathsf{f}$. In particular, $I_h \cup J_h \neq \varnothing$, for $(\pi_1 \circ h) \in \mathrm{Hom}(\mathcal{A}, \mathcal{BD}_2)$. Put $b \triangleq \bigwedge^{\mathfrak{A}}_{k \in n} (a_k \vee^{\mathfrak{A}} \sim^{\mathfrak{A}} a_k)$. Then, by (3.17), we have $\sim^{\mathfrak{A}} b \leqslant^{\mathfrak{A}} b$. Next, for every $k \in (I_h \cup J_h)$, set:

$$d_k \triangleq \begin{cases} a_k & \text{if } k \in I_h, \\ \sim^{\mathfrak{A}} a_k & \text{otherwise.} \end{cases}$$

Then, for each $g \in \mathrm{Hom}(\mathcal{A}, \mathcal{DM}_2)$, there is some $k \in (I_h \cup J_h)$ such that $gd_k = \mathsf{t}$. Finally, put $c \triangleq \bigvee^{\mathfrak{A}}_{k \in J_h \cup J_h} d_k$. Then, for each $g \in \mathrm{Hom}(\mathcal{A}, \mathcal{DM}_2)$, $gc = \mathsf{t}$. Moreover, for each $k \in (I_h \cup J_h)$, $hd_k \leqslant \langle \mathsf{f}, \mathsf{t} \rangle$, so $hc \leqslant \langle \mathsf{f}, \mathsf{t} \rangle$, in which case $\sim^{\mathfrak{A}} c \nleqslant^{\mathfrak{A}} c$. Further, take any $m \in \{\mathsf{n}, \mathsf{b}\}$ and any $f \in \mathrm{Hom}(\mathcal{A}, \mathcal{DM}_m)$. Consider the following two complementary cases:

1. $f[A] \subseteq \{\mathsf{f}, \mathsf{t}\}$.
   Then, $f \in \mathrm{Hom}(\mathcal{A}, \mathcal{DM}_2)$, for $\mathcal{DM}_2$ is a substructure of $\mathcal{DM}_m$, in which case $fc = \mathsf{t}$, and so $f(b \wedge^{\mathfrak{A}} \sim^{\mathfrak{A}} c) \leqslant f(\sim^{\mathfrak{A}} b \vee^{\mathfrak{A}} c)$.

2. $f[A] \nsubseteq \{\mathsf{f}, \mathsf{t}\}$.
   As the set $\{\mathsf{f}, \mathsf{t}\}$ forms a substructure of $\mathcal{DM}_m$, there must be some $k \in n$ such that $fa_k = m$, in which case $fb = m$, and so $f \sim^{\mathfrak{A}} b = m$ as well. Hence, $f(b \wedge^{\mathfrak{A}} \sim^{\mathfrak{A}} c) \leqslant f(\sim^{\mathfrak{A}} b \vee^{\mathfrak{A}} c)$.

Thus, in any case, $f(b \wedge^{\mathfrak{A}} \sim^{\mathfrak{A}} c) \leqslant f(\sim^{\mathfrak{A}} b \vee^{\mathfrak{A}} c)$. Therefore, by Corollary 3.16, $(b \wedge^{\mathfrak{A}} \sim^{\mathfrak{A}} c) \leqslant^{\mathfrak{A}} (\sim^{\mathfrak{A}} b \vee^{\mathfrak{A}} c)$. This, contradicts to the quasi-identity (3.28). In this way, there is some $g_h \in \mathrm{Hom}(\mathcal{A}, \mathcal{DM}_2)$ such that, for every $i \in I_h$, $g_h a_i = \mathsf{f}$, while, for every $j \in J_h$, $g_h a_j = \mathsf{t}$. Let $p_{1,2} : (DM_l \times DM_2) \times DM_2 \to (DM_l \times DM_2) \times DM_2, \langle a, b, c \rangle \mapsto \langle a, c, b \rangle$. This is clearly an automorphism of $(\mathcal{C} \times \mathcal{DM}_2) \times \mathcal{DM}_2$. Hence, $e \triangleq p_{1,2} \circ (\Delta_{4_l} \times \Delta_{DM_2})$ is an embedding of $(\mathcal{B} \downarrow 4) \times \mathcal{DM}_2$ into $(\mathcal{C} \times \mathcal{DM}_2) \times \mathcal{DM}_2$. Moreover, for each $k \in n$, $(h \times g_h)(a_k) \in e[4_l \times DM_2]$. Hence, as the set $e[4_l \times DM_2]$ forms a subalgebra of $(\mathfrak{C} \times \mathfrak{DM}_2) \times \mathfrak{DM}_2$, $(h \times g_h)[A] \subseteq e[4_l \times DM_2]$. Therefore, $e^{-1} \circ (h \times g_h) \in \mathrm{Hom}(\mathcal{A}, (\mathcal{B} \downarrow 4) \times \mathcal{DM}_2)$. Consider any $r \in (L + \approx)$. Then,

$$(e^{-1} \circ (h \times g_h))^{-1}[r^{((\mathcal{B}\downarrow 4)\times\mathcal{DM}_2))+\approx}] = (h \times g_h)^{-1}[e[r^{((\mathcal{B}\downarrow 4)\times\mathcal{DM}_2))+\approx}]] =$$
$$(h \times g_h)^{-1}[r^{(\mathcal{B}\times\mathcal{DM}_2)+\approx}] = h^{-1}[r^{\mathcal{B}+\approx}] \cap (g_h)^{-1}[r^{\mathcal{DM}_2+\approx}] \subseteq h^{-1}[r^{\mathcal{B}+\approx}].$$

Finally, since $\mathcal{DM}_2$ is a substructure of $\mathcal{C}$, $\Delta_{DM_2} \times \Delta_{DM_2}$ is an embedding of $\mathcal{DM}_2$ into $\mathcal{B} \downarrow 4$. Hence, $(\mathcal{B} \downarrow 4) \times \mathcal{DM}_2 \in \mathbf{I}(\mathcal{B} \downarrow 4)$. Thus, $\mathcal{A} \in \mathbf{I}(\{(\mathcal{B} \downarrow 4) \times \mathcal{DM}_2 | \mathcal{B} \in \mathsf{K}\} \subseteq \mathbf{I}(\mathsf{K} \downarrow 4)$, as required. ∎

By Lemma 3.45, Propositions 3.31, 3.32, 3.33, 3.44(i) and Corollary 3.35, we respectively obtain:

Proposition 3.46. $\mathsf{RSQKL} = \mathbf{I}(\{(\mathcal{DM}_\mathsf{b} \times \mathcal{DM}_2) \upharpoonright 4_\mathsf{b}, (\mathcal{DM}_\mathsf{n} \times \mathcal{DM}_2) \upharpoonright 4_\mathsf{n}\})$.

Proposition 3.47. $\mathsf{RNPSQKL} = \mathbf{I}((\mathcal{DM}_\mathsf{n} \times \mathcal{DM}_2) \upharpoonright 4_\mathsf{n})$.

Proposition 3.48. $\mathsf{RNCSQKL} = \mathbf{I}((\mathcal{DM}_\mathsf{b} \times \mathcal{DM}_2) \upharpoonright 4_\mathsf{b})$.

Corollary 3.49. (cf. Proposition 4.7 of [27]) $\mathsf{RPKL} = \mathbf{I}(\langle(\mathfrak{DM}_\mathsf{b}, \leqslant^{\mathfrak{DM}_\mathsf{b}}\rangle \times \mathcal{DM}_2) \upharpoonright 4_\mathsf{b})$.

3.2.4 Structurally complete subprevarieties

During this section, we assume that F is constant-free.

Theorem 3.50. NIPDML *is structurally complete, while:*

$$\mathbf{V}(\mathsf{NIPDML}) = \mathbf{V}(\mathsf{SQDML}).$$

Proof. By Corollary 3.34, $\langle \mathfrak{DM}_4, \leqslant^{\mathfrak{DM}_4}\rangle \times \mathcal{DM}_2 \in \mathbf{V}(\mathsf{NIPDML})$. On the other hand, π_0 is a surjective homomorphism from $\langle \mathfrak{DM}_4, \leqslant^{\mathfrak{DM}_4}\rangle \times \mathcal{DM}_2$ onto $\mathcal{DM}_4$, in which case $\mathcal{DM}_4 \in \mathbf{V}(\mathsf{NIPDML})$, so, by Theorem 3.15,

$$\mathbf{V}(\mathsf{NIPDML}) = \mathbf{V}(\mathsf{SQDML}).$$

Finally, consider any prevariety $\mathsf{P} \subseteq \mathsf{NIPDML}$. Assume $\mathbf{V}(\mathsf{P}) = \mathbf{V}(\mathsf{NIPDML}) = \mathbf{V}(\mathsf{SQDML})$. Then, by Corollary 3.23, $\mathsf{P} \not\subseteq \mathsf{PKL}$. Hence, by Lemma 2.10 together with Lemma 4.10 of [27], $\langle \mathfrak{DM}_4, \leqslant^{\mathfrak{DM}_4}\rangle \times \mathcal{DM}_2 \in \mathsf{P}$, so, by Corollary 3.34, $\mathsf{P} = \mathsf{NIPDML}$, as required. ∎

Lemma 3.51. *Let* $a \in \{\mathsf{n}, \mathsf{b}\}$, P *a subprevariety of* $\mathsf{RSQKL} \cap \mathbf{I}(\mathcal{DM}_a)$ *such that* $\mathsf{P} \nsubseteq \mathsf{SQBL}$. *Then,* $((\mathcal{DM}_a \times \mathcal{DM}_2) \upharpoonright 4_a) \in \mathsf{P}$.

Proof. By Remarks 3.20, 3.22 together with the proof of Case 3 of Theorem 4.8 of [27]. ■

Lemma 3.52. *Let* $a \in \{\mathsf{n}, \mathsf{b}\}$. *Then,* $\mathbf{V}(\mathsf{RSQKL} \cap \mathbf{I}(\mathcal{DM}_a)) = \mathbf{V}(\mathcal{DM}_a)$

Proof. By Propositions 3.47 and 3.48, $((\mathcal{DM}_a \times \mathcal{DM}_2) \upharpoonright 4_a) \in \mathbf{V}(\mathsf{RSQKL} \cap \mathbf{I}(\mathcal{DM}_a)$. Moreover, π_0 is a surjective homomorphism from $(\mathcal{DM}_a \times \mathcal{DM}_2) \upharpoonright 4_a$ onto $\mathcal{DM}_a$, so $\mathcal{DM}_a \in \mathbf{V}(\mathsf{RSQKL} \cap \mathbf{I}(\mathcal{DM}_a)$. This completes the argument. ■

By Corollaries 3.23, 3.19, 3.21, Propositions 3.47, 3.48, Lemmas 3.51 and 3.52, we immediately obtain the following two theorems:

Theorem 3.53. $\mathsf{RNPSQKL}$ *is structurally complete, while:*

$$\mathbf{V}(\mathsf{RNPSQKL}) = \mathbf{V}(\mathsf{NPSQKL}).$$

Theorem 3.54. $\mathsf{RNCSQKL}$ *is structurally complete, while:*

$$\mathbf{V}(\mathsf{RNCSQKL}) = \mathbf{V}(\mathsf{NCSQKL}).$$

Finally, by Remark 3.18 together with Theorem 4.8 of [27], we get:

Theorem 3.55. SQBL *is structurally complete.*

3.2.5 Summary

Thus, we have found all axiomatic extensions of the multiple-conclusion sequent calculus associated with $\mathcal{M}_4$ according to [31] that had been proposed and studied earlier in [22], [24], [25] (cf. [35]). In this connection, we discuss the following four consistent proper ones.

First, the axiomatic extension corresponding to SQKL is relatively axiomatized by the single axiom:

$$x, {\sim}x \rightarrowtail y, {\sim}y.$$

Second, the axiomatic extension corresponding to NPSQDML is relatively axiomatized by the single axiom:

$$x, {\sim}x \rightarrowtail y. \tag{3.29}$$

It is actuaaly associated with $\mathcal{M}_{\mathsf{n}} \triangleq \mathcal{M}_4 \upharpoonright \{\mathsf{f}, \mathsf{t}, \mathsf{n}\}$, defining Kleene's three-valued logic [13], according to [31]. Moreover, in view of Proposition 2.9 of [34], Remark 3.20, Proposition 2.4, Theorem 1.4(iii)⇔(ii), it is algebraizable with equivalent purely-algebraic semantics KL (cf. [32], [35]). In this way, in view of Corollary 1.8, the work [27] (in the constant-free case alone) provides the lattice of all extensions of the axiomatic extension under consideration together with their relative axiomatizations and finitely-generated equivalent purely-algebraic semantics. We highlight just one of them here. This corresponds to RNPSQKL and is relatively axiomatized by the single rule:

$$\frac{{\sim}x \rightarrowtail x \quad x, {\sim}y \rightarrowtail {\sim}x, y}{{\sim}y \rightarrowtail y} \tag{3.30}$$

It is constituted exactly by all rules admissible in the axiomatic extension under consideration.

Third, the axiomatic extension corresponding to NCSQDML is relatively axiomatized by the single axiom:

$$y \rightarrowtail x, {\sim}x. \tag{3.31}$$

It is actuaaly associated with $\mathcal{M}_{\mathsf{b}} \triangleq \mathcal{M}_4 \upharpoonright \{\mathsf{f}, \mathsf{t}, \mathsf{b}\}$, defining the *logic of paradox* [18] (cf. [23], [28]), according to [31]. Moreover, in view of Proposition 2.9 of [34], Remark 3.22, Proposition 2.4, Theorem 1.4(ii)⇔(iii), it is algebraizable with equivalent purely-algebraic semantics KL (cf. [32], [35]). In this way, in view of Corollary 1.8, the work [27] (in the constant-free case alone) provides the lattice of all extensions of the axiomatic extension under consideration together with their relative axiomatizations and finitely-generated equivalent purely-algebraic semantics. We highlight just one of them here. This corresponds to RNCSQKL and is relatively axiomatized by the single rule (3.30). It is constituted exactly by all rules admissible in the axiomatic extension under consideration.

Fourth, the axiomatic extension corresponding to SQBL is relatively axiomatized by the both axioms (3.29) and (3.31). It is actually associated with $\mathcal{M}_2 \triangleq \mathcal{M}_4 \upharpoonright \{\mathsf{f}, \mathsf{t}\}$ (with equality determinant $\{x_0\}$), defining the classical logic, according to [31] and is deductively equivalent to Gentzen's LK [8]. Moreover, in view of Proposition 2.9 of [34], Remark 3.18, Proposition 2.4, Theorem 1.4(ii)⇔(iii), it is algebraizable with equivalent purely-algebraic semantics BL (cf. [24], [32], [35]). In this way, in view of Corollary 1.8, the work [27] (for the cases both with and without constants, in view of Remark 3.12) implies that the axiomatic extension under consideration has no proper consistent extension.

After all, the extension of the sequent calculus mentioned above corresponding to PDML is relatively axiomatized by the single *Weak Contraposition* rule (cf. Proposition 3.24):

$$\frac{x \rightarrowtail y}{\sim y \rightarrowtail \sim x}. \tag{3.32}$$

It is algebraizable with equivalent purely-algebraic semantics DML (cf. [24], [25] and Remark 2.11). In this way, the work [27] (in the constant-free case alone) had provided us with the lattice of its extensions. We highlight just one of them here. This corresponds to NIPDML and is relatively axiomatized by the single rule:

$$\frac{x \rightarrowtail \sim x \qquad \sim x \rightarrowtail x}{\rightarrowtail} \tag{3.33}$$

It is constituted exactly by all rules admissible in the initial sequent calculus.

3.3 The classical expansion of Belnap's four-valued logic

During this section, we put $F \triangleq \{\wedge, \vee, \bot, \top, \neg, \sim\}$ and $L' \triangleq \langle F, \{D\}\rangle$ (the sentential case). In addition, we deal with the following secondary binary connectives:

$$\begin{aligned}(x \leftrightarrow y) &\triangleq ((\neg x \vee y) \wedge (\neg y \vee x)),\\ (x \equiv y) &\triangleq ((x \leftrightarrow y) \wedge (\sim x \leftrightarrow \sim y)).\end{aligned}$$

Recall that a *Boolean De Morgan algebra* (cf. [25], [29]) is any F-algebra, whose $(F \setminus \{\sim\})$-reduct is a Boolean algebra and whose $F \setminus \{\neg\}$-reduct is a De Morgan algebra, the variety of all them being denoted by BDMA. Boolean De Morgan algebras also satisfy the following identity:

$$\sim\neg x \approx \neg\sim x. \tag{3.34}$$

A *filtered Boolean De Morgan algebra* is any L'-structure (i.e., F-matrix), whose underlying algebra is a Boolean De Morgan algebra, and that satisfies the following four additional quasi-identities:

$$D(x \equiv y) \rightarrow (x \approx y), \tag{3.35}$$

$$(D(x) \& D(y)) \rightarrow D(x \wedge y), \tag{3.36}$$

$$D(x \wedge y) \rightarrow D(x), \tag{3.37}$$

$$D(\top), \tag{3.38}$$

the quasivariety of all them being denoted by FBDMA.

Put $\mathcal{BD}_4 \triangleq \langle \mathfrak{BD}_4, (\pi_0)^{-1}[\{1\}] \rangle$, where $\mathfrak{BD}_4$ is the expansion of $(\mathfrak{B}_2)^2$, where $\mathfrak{B}_2$ is, in its turn, the standard two-element Boolean algebra with carrier 2 and $0 < 1$, to F given by $\sim^{\mathfrak{BD}_4} \triangleq \sim^{\mathfrak{DM}_4}$. This matrix with the same equality determinant Υ (cf. Section 3.2) defines the expansion of Belnap' four-valued logic by the classical negation (cf. [25]). Put $\mathcal{BD}_2 \triangleq \mathcal{BD}_4 \upharpoonright \{\mathsf{f}, \mathsf{t}\}$.

Remark 3.56. $\mathcal{BD}_4$ and $(\mathcal{BD}_4)^{[\{1\},\{1\}]}$ are relationally rationally equivalent with respect to the purely relational (L', L)-translation τ, defined by

$$\tau(D) \triangleq \{\top Q x_0\},$$

and (L, L')-translation ρ, defined by $\rho(Q) \triangleq \{D(\neg x_0 \vee x_1)\}$. ■

Lemma 3.57. *Let $\mathfrak{A}$ be a Boolean De Morgan algebra and X a proper prime non-empty filter of it. Then, we have the $h^{\mathfrak{A}}_X \in \mathrm{Hom}(\mathfrak{A}, \mathfrak{BD}_4)$ defined by $h^{\mathfrak{A}}_X \triangleq \chi^X_A \times (\chi^{A \setminus X}_A \circ \sim^{\mathfrak{A}})$, in which case $X = (h^{\mathfrak{A}}_X)^{-1}[D^{\mathcal{BD}_4}]$.*

Proof. By (3.1), (3.6), (3.34) together with Lemma 3.14. ■

By (3.6), the identity $(x \equiv x) = \top$ is valid in Boolean De Morgan algebras, so, by (3.38), the following identity is valid in filtered Boolean De Morgan algebras:

$$D(x \equiv x). \tag{3.39}$$

Put $\varepsilon \triangleq \{D(x_0 \equiv x_1)\}$.

Proposition 3.58. FBDMA $= \mathbf{I}(\mathcal{BD}_4)$.

Proof. Clearly, $\mathfrak{BD}_4$ is a Boolean De Morgan algebra. Moreover, $D^{\mathcal{BD}_4}$ is a prime non-empty filter of $\mathfrak{BD}_4$. In particular, (3.36), (3.37) and (3.38) are true in $\mathcal{BD}_4$, while $(a \in D^{\mathcal{BD}_4} \Rightarrow b \in D^{\mathcal{BD}_4}) \Leftrightarrow (\neg^{\mathfrak{BD}_4} a \vee^{\mathfrak{BD}_4} b) \in D^{\mathcal{BD}_4}$, for all $a, b \in BD_4$. Hence, since Υ is an equality determinant for $\mathcal{BD}_4$, (3.35) is true in it. Thus, $\mathcal{BD}_4 \in$ FBDMA, so $\mathbf{I}(\mathcal{BD}_4) \subseteq$ FBDMA. For proving the converse inclusion, consider any $\mathcal{A} \in$ FBDMA. Take an arbitrary $a \in A$ such that $a \notin D^{\mathcal{A}}$. Then, by the Prime Ideal Theorem for distributive lattices, there is a prime filter X of $\mathfrak{A}$ such that $D^{\mathcal{A}} \subseteq X$, whereas $a \notin X$, in which case X is proper and non-empty, in view of (3.38). Then, by Lemma 3.57, $h^{\mathfrak{A}}_X \in \mathrm{Hom}(\mathcal{A}, \mathcal{BD}_4)$, while $a \notin (h^{\mathfrak{A}}_X)^{-1}[D^{\mathcal{BD}_4}]$. Hence, by (3.35), (3.39) and Proposition 1.15, $\mathcal{A} \in \mathbf{I}(\mathcal{BD}_4)$, as required. ■

A filtered Boolean De Morgan algebra is said to be *truth-singular*, if it satisfies the following quasi-identity:[5]

$$D(x) \to (x \approx \top). \tag{3.40}$$

The quasivariety of all truth-singular filtered Boolean De Morgan algebras is denoted by $\mathsf{TSFBDMA}$. By the identity (3.38), we have:

Lemma 3.59. $\mathsf{TSFBDMA} = \{\langle \mathfrak{A}, \{\top^{\mathfrak{A}}\}\rangle | \mathfrak{A} \in \mathsf{BDMA}\}$.

Remark 3.60. Lemma 3.59 means that the quasivarieties BDMA and $\mathsf{TSFBDMA}$ are relationally rationally equivalent with respect to ι_F and the purely relational $(L', F + \approx)$-translation ϑ, defined by $\vartheta(D) \triangleq \{x_0 \approx \top\}$. ■

In this way, all the results proved in [29] for BDMA are properly transferred to $\mathsf{TSFBDMA}$. In particular, Proposition 3.1 of [29] and Lemma 3.59 yield:

Corollary 3.61. $\mathsf{TSFBDMA} = \mathbf{I}(\langle \mathfrak{BD}_4, \{\top^{\mathfrak{BD}_4}\}\rangle)$.

Note that, by (3.38), any singular filtered Boolean De Morgan algebra is trivial. The quasivariety of all singular filtered Boolean De Morgan algebras is denoted by SFBDMA. Clearly, $\mathsf{SFBDMA} = \mathbf{I}(\varnothing) \subseteq \mathsf{TSFBDMA}$.

The quasivariety of all non-idempotent (and truth-singular) filtered Boolean De Morgan algebras is denoted by $\mathsf{NIFBDMA}$ (resp., by $\mathsf{NITSFBDMA}$). Notice that:

$$\mathcal{BD}_2 = \langle \mathfrak{BD}_2, \{\top^{\mathfrak{BD}_2}\}\rangle. \tag{3.41}$$

Then, by Proposition 3.6 of [29] and Lemma 3.59, we immediately get:[6]

Corollary 3.62. $\mathsf{NITSFBDMA} = \mathbf{I}(\langle \mathfrak{BD}_4, \{\top^{\mathfrak{BD}_4}\}\rangle \times \mathcal{BD}_2)$.

A filtered Boolean De Morgan algebra is said to be *classical* if it satisfies the following identity:

$$\sim x \approx \neg x. \tag{3.42}$$

The quasivariety of all classical filtered Boolean De Morgan algebras is denoted by CFBDMA. In view of (3.35), $\mathsf{CFBDMA} \subseteq \mathsf{TSFBDMA}$. Then, Corollary 3.4 of [29], (3.41) and Lemma 3.59 immediately imply:

[5]Within the context of General Logic, the unary relation symbol D is treated as "truth predicate". This explains the terminology adopted here.

[6]It equally follows from Proposition 3.61 and Corollary 3.29.

Corollary 3.63. $\mathsf{CFBDMA} = \mathbf{I}(\mathcal{BD}_2)$.

Remark 3.64. In this way, the quasivariety CFBDMA is rationally equivalent to the variety of Boolean algebras vith respect to the $(L', (F\setminus\{\sim\})+\approx)$-translation ϱ, being the extension of $\iota_{F\setminus\{\sim\}}$ to L', defined by $\varrho(D) \triangleq \{x_0 \approx \top\}$ and $\varrho(\sim) \triangleq \neg x_0$, and $\iota_{F\setminus\{\sim\}}$, for $\mathcal{BD}_2$ and $\mathfrak{B}_2$ are so (cf. Theorem 1.4(iv)⇒(iii)). ∎

As $\mathcal{BD}_4$ has no singular substructure, By Corollaries 3.29, 3.61, Lemma 3.36 and Proposition 3.58, we immediately obtain the following two assertions:

Proposition 3.65. $\mathsf{NIFBDMA} = \mathbf{I}(\mathcal{BD}_4 \times \mathcal{BD}_2)$.

Proposition 3.66.

$$\mathsf{NIFBDMA} \cup \mathsf{TSFBDMA} = \mathbf{I}(\{\mathcal{BD}_4 \times \mathcal{BD}_2, \langle \mathfrak{BD}_4, \{\top^{\mathfrak{BD}_4}\}\rangle\})$$

is the subquasivariety of FBDMA *relatively axiomatized by the single quasi-identity:*

$$(\sim x \approx x)\&D(y) \to (y \approx \top). \tag{3.43}$$

A filtered Boolean De Morgan algebra is said to be *non-paraconsistent* if it satisfies the following quasi-identity:

$$D(\sim x \wedge x) \to (x \approx y), \tag{3.44}$$

the quasivariety of all them being denoted by $\mathsf{NPFBDMA}$.

Lemma 3.67. *Let $\mathcal{A}$ be a non-singular finitely-generated non-paraconsistent filtered Boolean De Morgan algebra. Then,* $\mathrm{Hom}(\mathcal{A}, \langle \mathfrak{BD}_4, \{\top^{\mathfrak{BD}_4}\}\rangle) \neq \varnothing$.

Proof. By contradiction. For suppose $\mathrm{Hom}(\mathcal{A}, \langle \mathfrak{BD}_4, \{\top^{\mathfrak{BD}_4}\}\rangle) = \varnothing$. As $\mathcal{A}$ is finitely-generated, while $\mathcal{BD}_4$ is finite, the set $\mathrm{Hom}(\mathcal{A}, \mathcal{BD}_4)$ is finite. Moreover, since $\mathcal{A}$ is not singular, by Proposition 3.58, $\mathrm{Hom}(\mathcal{A}, \mathcal{BD}_4) \neq \varnothing$. Consider any $g, f \in \mathrm{Hom}(\mathcal{A}, \mathcal{BD}_4)$. Let us prove, by contradiction, that, there is some $a \in A$ such that $ga = \mathsf{b}$, while $fa \leqslant \mathsf{b}$. For assume $\{\langle \mathsf{b}, \mathsf{b}\rangle, \langle \mathsf{b}, \mathsf{f}\rangle\} \cap (g \times f)[A] = \varnothing$. Then, $\{\langle \mathsf{n}, \mathsf{n}\rangle, \langle \mathsf{b}, \mathsf{t}\rangle, \langle \mathsf{n}, \mathsf{f}\rangle, \langle \mathsf{n}, \mathsf{t}\rangle\} \cap (g \times f)[A] = \varnothing$. Consider the following two complementary cases:

1. $\langle \mathsf{b}, \mathsf{n}\rangle \notin (g \times f)[A]$.
   Then, $\mathsf{b} \notin g[A]$, and so $\mathsf{n} \notin g[A]$, in which case:

   $$g \in \mathrm{Hom}(\mathcal{A}, \mathcal{BD}_2) \subseteq \mathrm{Hom}(\mathcal{A}, \langle \mathfrak{BD}_4, \{\top^{\mathfrak{BD}_4}\}\rangle),$$

   for $\mathcal{BD}_2$ is a substructure of $\langle \mathfrak{BD}_4, \{\top^{\mathfrak{BD}_4}\}\rangle$, in view of (3.41).

2. $\langle \mathsf{b}, \mathsf{n} \rangle \in (g \times f)[A]$.
Then, $\langle \mathsf{n}, \mathsf{b} \rangle \in (g \times f)[A]$, in which case:

$$\{\langle \mathsf{f}, \mathsf{t} \rangle, \langle \mathsf{t}, \mathsf{f} \rangle, \langle \mathsf{f}, \mathsf{n} \rangle, \langle \mathsf{t}, \mathsf{n} \rangle, \langle \mathsf{f}, \mathsf{b} \rangle, \langle \mathsf{t}, \mathsf{b} \rangle\} \cap (g \times f)[A] = \varnothing.$$

Therefore, $(g \times f)[A] = B \triangleq \{\langle \mathsf{f}, \mathsf{f} \rangle, \langle \mathsf{t}, \mathsf{t} \rangle, \langle \mathsf{b}, \mathsf{n} \rangle, \langle \mathsf{n}, \mathsf{b} \rangle\}$, in which case:

$$(g \times f) \in \mathrm{Hom}(\mathcal{A}, (\mathcal{BD}_4)^2 \upharpoonright B).$$

Moreover, π_0 is an isomorphism from $(\mathcal{BD}_4)^2 \upharpoonright B$ onto $\langle \mathfrak{BD}_4, \{\top^{\mathfrak{BD}_4}\} \rangle$, so:

$$\pi_0 \circ (g \times f) \in \mathrm{Hom}(\mathcal{A}, \langle \mathfrak{BD}_4, \{\top^{\mathfrak{BD}_4}\} \rangle).$$

Both cases contradict the assumption that $\mathrm{Hom}(\mathcal{A}, \langle \mathfrak{BD}_4, \{\top^{\mathfrak{BD}_4}\} \rangle) = \varnothing$. Thus, there is some $a^g_f \in A$ such that $ga^g_f = \mathsf{b}$, while $fa^g_f \leqslant \mathsf{b}$. Next, consider any $h \in \mathrm{Hom}(\mathcal{A}, \mathcal{BD}_4)$. Put $a^h \triangleq \bigwedge^{\mathfrak{A}} \{a^h_e | e \in \mathrm{Hom}(\mathcal{A}, \mathcal{BD}_4)\}$. Then, $ha^h = \mathsf{b}$, while $ea^h \leqslant \mathsf{b}$, for every $e \in \mathrm{Hom}(\mathcal{A}, \mathcal{BD}_4)$. Finally, set $a \triangleq \bigvee^{\mathfrak{A}} \{a^h | h \in \mathrm{Hom}(\mathcal{A}, \mathcal{BD}_4)\}$. Then, for each $h \in \mathrm{Hom}(\mathcal{A}, \mathcal{BD}_4)$, $ha = \mathsf{b}$, in which case $h({\sim}^{\mathfrak{A}} a \wedge^{\mathfrak{A}} a) = \mathsf{b} \in D^{\mathcal{BD}_4}$. Therefore, by Proposition 3.58, ${\sim}^{\mathfrak{A}} a \wedge^{\mathfrak{A}} a \in D^{\mathcal{A}}$. In view of (3.44), this contradicts to the assumption that $\mathcal{A}$ is not singular. This contradiction completes the argument. ∎

Proposition 3.68. $\mathsf{NPFBDMA} = \mathbf{I}(\mathcal{BD}_4 \times \langle \mathfrak{BD}_4, \{\top^{\mathfrak{BD}_4}\} \rangle)$.

Proof. Clearly, $\mathcal{BD}_4 \times \langle \mathfrak{BD}_4, \{\top^{\mathfrak{BD}_4}\} \rangle$ is non-paraconsistent. This yields the inclusion from right to left. Since the structure involved is finite, the prevariety generated by it is a quasivariety, so it suffices to prove the inverse inclusion for finitely-generated structures alone. For take any finitely-generated non-paraconsistent filtered Boolean De Morgan algebra $\mathcal{A}$. Consider an arbitrary $a \in A \setminus D^{\mathcal{A}}$, in which case $a \neq \top^{\mathfrak{A}}$, in view of the identity (3.38). In that case, $\mathcal{A}$ is not singular, so, in view of Lemma 3.67, there is some $g \in \mathrm{Hom}(\mathcal{A}, \langle \mathfrak{BD}_4, \{\top^{\mathfrak{BD}_4}\} \rangle)$. Moreover, by Proposition 3.58, there is some $h \in \mathrm{Hom}(\mathcal{A}, \mathcal{BD}_4)$ such that $ha \notin D^{\mathcal{BD}_4}$. Then, $(h \times g) \in \mathrm{Hom}(\mathcal{A}, \mathcal{BD}_4 \times \langle \mathfrak{BD}_4, \{\top^{\mathfrak{BD}_4}\} \rangle)$, while $(h \times g)(a) \notin D^{\mathcal{BD}_4 \times \langle \mathfrak{BD}_4, \{\top^{\mathfrak{BD}_4}\} \rangle}$. Thus, by (3.35), (3.39) and Proposition 1.15, $\mathcal{A} \in \mathbf{I}(\mathcal{BD}_4 \times \langle \mathfrak{BD}_4, \{\top^{\mathfrak{BD}_4}\} \rangle)$, as required. ∎

By (3.35), (3.39) and Proposition 1.17, we immediately obtain:

Lemma 3.69. *Any filtered Boolean De Morgan algebra is simple.*

Theorem 3.70. *Implicational classes of filtered Boolean De Morgan lattices form the eight-element non-chain distributive lattice depicted at Figure 3.2,* FBDMA, CFBDMA *and* SFBDMA *being exactly all relative subvarieties of* FBDMA.

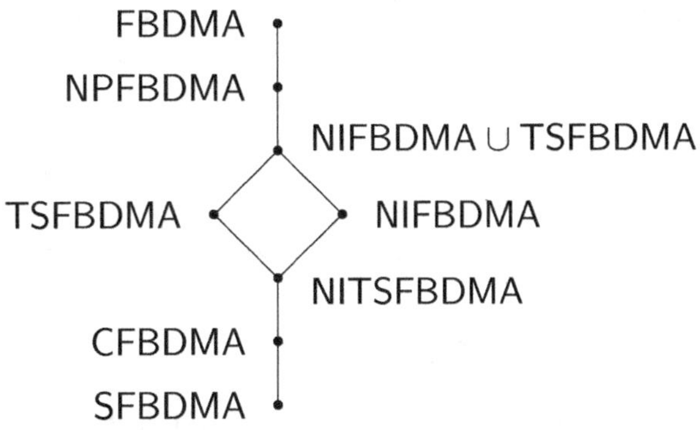

Figure 3.2: The lattice of implicational classes of filtered Boolean De Morgan algebras.

Proof. The case of implicational classes of truth-singular filtered Boolean De Morgan algebras is due to Theorem 3.7 of [29] and Lemma 3.59. It only remains to analyze the opposite case.

First, note that (3.44) is not true in $\mathcal{BD}_4$ under the assignment $[x/\mathsf{b}, y/\mathsf{t}]$, so NPFBDMA $\subsetneq$ FBDMA. Next, any filtered Boolean De Morgan algebra satisfying the identity $\top \approx \bot$ is singular, so TSFBDMA $\subseteq$ NPFBDMA. Moreover, by (3.35), (3.36) and (3.37), the following quasi-identity is true in FBDMA:

$$D(x \wedge \sim x) \to (x \approx \sim x), \tag{3.45}$$

so NIFBDMA $\subseteq$ NPFBDMA. Furthermore, (3.43) is not true in $\mathcal{BD}_4 \times \langle \mathfrak{BD}_4, \{\top^{\mathfrak{BD}_4}\}\rangle$ under the assignment $[x/\langle \mathsf{b}, \mathsf{b}\rangle, y/\langle \mathsf{b}, \mathsf{t}\rangle]$. Thus, by Propositions 3.66 and 3.68, we get NIFBDMA$\cup$TSFBDMA $\subsetneq$ NPFBDMA. Further, $\mathcal{BD}_4 \times \mathcal{BD}_2$ is not truth-singular, while $\langle \mathfrak{BD}_4, \{\top^{\mathfrak{BD}_4}\}\rangle$ is not non-idempotent, so, by Corollary 3.61 and Proposition 3.65, NIFBDMA $\not\subseteq$ TSFBDMA $\not\subseteq$ NIFBDMA. In this way, the eight prevarieties of filtered Boolean De Morgan algebras involved do form the lattice depicted at Figure 3.2. It only remains to argue that there is no more prevariety of filtered Boolean De Morgan algebras. For take any prevariety P $\not\subseteq$ TSFBDMA of filtered Boolean De Morgan algebras. Consider the following four exhaustive cases:

1. P $\not\subseteq$ NPFBDMA.
   Then, there is a filtered Boolean De Morgan algebra $\mathcal{A} \in$ P not satisfying (3.44), in which case $\mathcal{A}$ is not singular, while there is some $a \in A$ such that $a \wedge^{\mathfrak{A}} \sim^{\mathfrak{A}} a \in D^{\mathcal{A}}$, in which case, by (3.45), $a = \sim^{\mathfrak{A}} a$, and so $\neg^{\mathfrak{A}} a =$

$\sim^{\mathfrak{A}}\neg^{\mathfrak{A}}a$. Since $\mathcal{A}$ is not singular, the elements a, $\neg^{\mathfrak{A}}a$, $\bot^{\mathfrak{A}}$ and $\top^{\mathfrak{A}}$ are all pairwise distinct, while $\{\neg^{\mathfrak{A}}a, \bot^{\mathfrak{A}}\} \cap D^{\mathcal{A}} = \varnothing$, whereas $\{a, \top^{\mathfrak{A}}\} \subseteq D^{\mathcal{A}}$. Therefore, we have the embedding e of $\mathcal{BD}_4$ into $\mathcal{A}$ defined by:

$$\begin{aligned} e\mathsf{b} &\triangleq a, \\ e\mathsf{n} &\triangleq \neg^{\mathfrak{A}}a, \\ e\mathsf{t} &\triangleq \top^{\mathfrak{A}}, \\ e\mathsf{f} &\triangleq \bot^{\mathfrak{A}}. \end{aligned}$$

Hence, $\mathcal{BD}_4 \in \mathsf{P}$, so, by Proposition 3.58, $\mathsf{P} = \mathsf{FBDMA}$.

2. $\mathsf{P} \subseteq \mathsf{NPFBDMA}$ but $\mathsf{P} \nsubseteq \mathsf{NIFBDMA} \cup \mathsf{TSFBDMA}$.
Then, by Proposition 3.66, there are some $\mathcal{A} \in \mathsf{P}$, $c \in A$ and $b \in D^{\mathcal{A}}$ such that $\sim^{\mathfrak{A}}c = c$, while $b \neq \top^{\mathfrak{A}}$, in which case $\mathcal{A}$ is not singular. Put:

$$a \triangleq \begin{cases} c & \text{if } c \vee^{\mathfrak{A}} b \neq \top^{\mathfrak{A}}, \\ \neg^{\mathfrak{A}}c & \text{otherwise.} \end{cases}$$

Then, $\sim^{\mathfrak{A}}a = a$, in which case $\sim^{\mathfrak{A}}\neg^{\mathfrak{A}}a = \neg^{\mathfrak{A}}a$, while $a \vee^{\mathfrak{A}} b \neq \top^{\mathfrak{A}}$, whereas $a \vee^{\mathfrak{A}} b \in D^{\mathcal{A}}$. Moreover, since $\mathcal{A}$ is not singular, by (3.44) and (3.36), we have $\{a, \neg^{\mathfrak{A}}a, a \vee^{\mathfrak{A}} \neg^{\mathfrak{A}}b\} \cap D^{\mathcal{A}} = \varnothing$. Moreover, by (3.35), (3.36), (3.37) and (3.38), the following quasi-identity:

$$D(\neg{\sim}x \wedge x) \to (x \approx \top) \tag{3.46}$$

is true in FBDMA, and so in $\mathcal{A}$. Therefore, $\neg^{\mathfrak{A}}a \vee^{\mathfrak{A}} \neg^{\mathfrak{A}}{\sim}^{\mathfrak{A}}b \notin D^{\mathcal{A}}$. And what is more, the quasi-identity:

$$(x \approx {\sim}x)\&D(y)\&D(\neg x \vee {\sim}y) \to (\bot \approx \top)$$

is true in $\mathcal{BD}_4 \times \langle \mathfrak{BD}_4, \{\top^{\mathfrak{BD}_4}\}\rangle$, and so in $\mathcal{A}$, in view of Proposition 3.68. Therefore, $\neg^{\mathfrak{A}}a \vee^{\mathfrak{A}} {\sim}^{\mathfrak{A}}b \notin D^{\mathcal{A}}$. In this way, by (3.37) and (3.38), we have the strict homomorphism h from $\mathcal{BD}_4 \times \langle \mathfrak{BD}_4, \{\top^{\mathfrak{BD}_4}\}\rangle$ to $\mathcal{A}$

defined by:

$$
\begin{aligned}
h\langle \mathsf{f},\mathsf{f}\rangle &\triangleq \bot^{\mathfrak{A}},\\
h\langle \mathsf{t},\mathsf{t}\rangle &\triangleq \top^{\mathfrak{A}},\\
h\langle \mathsf{b},\mathsf{b}\rangle &\triangleq a,\\
h\langle \mathsf{n},\mathsf{n}\rangle &\triangleq \neg^{\mathfrak{A}}a,\\
h\langle \mathsf{b},\mathsf{t}\rangle &\triangleq a \vee^{\mathfrak{A}} b,\\
h\langle \mathsf{b},\mathsf{f}\rangle &\triangleq a \wedge^{\mathfrak{A}} {\sim}^{\mathfrak{A}}b,\\
h\langle \mathsf{n},\mathsf{f}\rangle &\triangleq \neg^{\mathfrak{A}}a \wedge^{\mathfrak{A}} \neg^{\mathfrak{A}}b,\\
h\langle \mathsf{n},\mathsf{t}\rangle &\triangleq \neg^{\mathfrak{A}}a \vee^{\mathfrak{A}} \neg^{\mathfrak{A}}{\sim}^{\mathfrak{A}}b,\\
h\langle \mathsf{f},\mathsf{n}\rangle &\triangleq \neg^{\mathfrak{A}}a \wedge^{\mathfrak{A}} b,\\
h\langle \mathsf{t},\mathsf{n}\rangle &\triangleq \neg^{\mathfrak{A}}a \vee^{\mathfrak{A}} {\sim}^{\mathfrak{A}}b,\\
h\langle \mathsf{f},\mathsf{b}\rangle &\triangleq a \wedge^{\mathfrak{A}} \neg^{\mathfrak{A}}{\sim}^{\mathfrak{A}}b,\\
h\langle \mathsf{t},\mathsf{b}\rangle &\triangleq a \vee^{\mathfrak{A}} \neg^{\mathfrak{A}}b,\\
h\langle \mathsf{t},\mathsf{f}\rangle &\triangleq (a \wedge^{\mathfrak{A}} {\sim}^{\mathfrak{A}}b) \vee^{\mathfrak{A}} (\neg^{\mathfrak{A}}a \wedge^{\mathfrak{A}} \neg^{\mathfrak{A}}b),\\
h\langle \mathsf{f},\mathsf{t}\rangle &\triangleq (a \vee^{\mathfrak{A}} b) \wedge^{\mathfrak{A}} (\neg^{\mathfrak{A}}a \vee^{\mathfrak{A}} \neg^{\mathfrak{A}}{\sim}^{\mathfrak{A}}b),\\
h\langle \mathsf{b},\mathsf{n}\rangle &\triangleq (a \vee^{\mathfrak{A}} b) \wedge^{\mathfrak{A}} (\neg^{\mathfrak{A}}a \vee^{\mathfrak{A}} {\sim}^{\mathfrak{A}}b),\\
h\langle \mathsf{n},\mathsf{b}\rangle &\triangleq (\neg^{\mathfrak{A}}a \wedge^{\mathfrak{A}} \neg^{\mathfrak{A}}b) \vee^{\mathfrak{A}} (a \wedge^{\mathfrak{A}} \neg^{\mathfrak{A}}{\sim}^{\mathfrak{A}}b).
\end{aligned}
$$

Then, $\ker h \in \mathrm{Con}(\mathcal{BD}_4 \times \langle \mathfrak{BD}_4, \{\top^{\mathfrak{BD}_4}\}\rangle)$, so, by Lemma 3.69, h is an embedding. Hence, $\mathcal{BD}_4 \times \langle \mathfrak{BD}_4, \{\top^{\mathfrak{BD}_4}\}\rangle \in \mathsf{P}$, so, by Proposition 3.68, $\mathsf{P} = \mathsf{NPFBDMA}$.

3. $\mathsf{P} \subseteq \mathsf{NIFBDMA}$.
   Take any non-truth-singular $\mathcal{A} \in \mathsf{P}$. Then, there is some $a \in D^{\mathcal{A}}$ such that $a \neq \top^{\mathfrak{A}}$, in which case $\mathcal{A}$ is not singular. Put $b \triangleq a \vee^{\mathfrak{A}} {\sim}^{\mathfrak{A}}a$, in which case $b \in D^{\mathcal{A}}$, while ${\sim}^{\mathfrak{A}}b \leqslant^{\mathfrak{A}} b$, whereas $b \neq \top^{\mathfrak{A}}$, for, otherwise, we would have ${\sim}^{\mathfrak{A}}a = \neg^{\mathfrak{A}}a$, in which case, by (3.46), we would get $a = \top^{\mathfrak{A}}$. Moreover, as $\mathcal{A}$ is both non-idempotent and non-singular, ${\sim}^{\mathfrak{A}}b \neq b$. Therefore, by (3.45), ${\sim}^{\mathfrak{A}}b \notin D^{\mathcal{A}}$, in which case, by (3.36), ${\sim}^{\mathfrak{A}}b \vee^{\mathcal{A}} \neg^{\mathfrak{A}}b \notin D^{\mathcal{A}}$, while, by (3.46), $\neg^{\mathfrak{A}}{\sim}^{\mathfrak{A}}b \notin D^{\mathcal{A}}$. Moreover, since $\mathcal{A}$ is non-singular, $\bot^{\mathfrak{A}} \notin D^{\mathcal{A}}$, so $\neg^{\mathfrak{A}}b \notin D^{\mathcal{A}}$. In this way, by (3.37) and (3.38), we have the

strict homomorphism h from $\mathcal{BD}_4 \times \mathcal{BD}_2$ to $\mathcal{A}$ defined by:

$$\begin{aligned}
h\langle \mathsf{f}, \mathsf{f} \rangle &\triangleq \bot^{\mathfrak{A}}, \\
h\langle \mathsf{t}, \mathsf{t} \rangle &\triangleq \top^{\mathfrak{A}}, \\
h\langle \mathsf{b}, \mathsf{t} \rangle &\triangleq b, \\
h\langle \mathsf{b}, \mathsf{f} \rangle &\triangleq \sim^{\mathfrak{A}} b, \\
h\langle \mathsf{n}, \mathsf{f} \rangle &\triangleq \neg^{\mathfrak{A}} b, \\
h\langle \mathsf{n}, \mathsf{t} \rangle &\triangleq \neg^{\mathfrak{A}} \sim^{\mathfrak{A}} b, \\
h\langle \mathsf{t}, \mathsf{f} \rangle &\triangleq \sim^{\mathfrak{A}} b \vee^{\mathfrak{A}} \neg^{\mathfrak{A}} b, \\
h\langle \mathsf{f}, \mathsf{t} \rangle &\triangleq b \wedge^{\mathfrak{A}} \neg^{\mathfrak{A}} \sim^{\mathfrak{A}} b.
\end{aligned}$$

Then, $\ker h \in \mathrm{Con}(\mathcal{BD}_4 \times \mathcal{BD}_2)$, so, by Lemma 3.69, h is an embedding. Therefore, $\mathcal{BD}_4 \times \mathcal{BD}_2 \in \mathsf{P}$, so, by Proposition 3.65, $\mathsf{P} = \mathsf{NIFBDMA}$.

4. $\mathsf{P} \subseteq \mathsf{TSFBDMA} \cup \mathsf{NIFBDMA}$ but $\mathsf{P} \not\subseteq \mathsf{NIFBDMA}$.
Then, $\mathsf{P} \cap \mathsf{NIFBDMA} \subseteq \mathsf{NIFBDMA}$ but $\mathsf{P} \cap \mathsf{NIFBDMA} \not\subseteq \mathsf{TSFBDMA}$, in which case, by Case 3, $\mathsf{P} \cap \mathsf{NIFBDMA} = \mathsf{NIFBDMA}$. Likewise, $\mathsf{P} \cap \mathsf{TSFBDMA} \subseteq \mathsf{TSFBDMA}$ but $\mathsf{P} \cap \mathsf{TSFBDMA} \not\subseteq \mathsf{NIFBDMA}$, in which case, by Theorem 3.7 of [29] and Lemma 3.59, $\mathsf{P} \cap \mathsf{TSFBDMA} = \mathsf{TSFBDMA}$. In this way, $\mathsf{P} = \mathsf{TSFBDMA} \cup \mathsf{NIFBDMA}$.

Thus, Theorem 3.7 of [29] and Lemma 3.59 complete the argument of that part of the theorem which concerns arbitrary subprevarieties of FBDMA. Finally, the part concerining relative subvarieties of FBDMA then follows immediately from (1.2) and Proposition 3.58, for $\mathfrak{BD}_2$ is the only subalgebra of $\mathfrak{BD}_4$. ■

As a consequence, we also get:

Corollary 3.71. NITSFBDMA *is structurally complete, while:*

$$\mathbf{V}(\mathsf{NITSFBDMA}) = \mathbf{V}(\mathsf{FBDMA}).$$

Thus, we have found all (in particular, axiomatic) extensions of the classical expansion of Belnap's four-valued logic together with their finite relative axiomatizations as well as their finite matrix semantics. (This equally concerns the relationally equivalent sequent calculus associated with $\mathcal{BD}_4$ according to [31] that had been proposed earlier in [25]; cf. Remark 3.56). In this connection, we should like to highlight the following three points.

First, NPFBDMA corresponds to the extension relatively axiomatized by the single *Ex Contradictione Quodlibet* rule:

$$x, \sim x \to y$$

Logics *not* satisfying this rule are referred to as *paraconsistent.* (This clarifies the term "non-paraconsistent" we have used with respect to members of NPFBDMA.) Since this is the least proper extension of the logic under consideration, we conclude that the latter is a *maximal paraconsistent logic* in the sense that it is paraconsistent, while no proper extension of it is. Such a behaviour has been observed for certain three-valued paraconsistent logics like P^1 [36] (cf. [21]) and the *logic of paradox* [18] (cf. [23] and [28]). In this way, the logic under consideration appears seemingly the first example of a *four-valued* maximal paraconsistent logic.

Second, the only proper consistent axiomatic extension is the logic of the F-matrix $\mathcal{BD}_2$ being the extension, corresponding to CFBDMA and relatively axiomatized by the single *Excluded Midle* axiom:

$$x \vee {\sim}x$$

It is functionally equivalent (in the sense of [34]) vith respect to ϱ_{f} and $\iota_{F\setminus\{\sim\}}$ to the standard version of the classical logic defined by the $(F \setminus \{\sim\})$-matrix $\langle \mathfrak{B}_2, \{1\}\rangle$ (cf. Remark 3.64). (This clarifies the term "classical" we have used with respect to members of CFBDMA.)

Third, the extension of the sequent calculus mentioned above corresponding to TSFBDMA is relatively axiomatized by (3.32), because $\neg^{\mathfrak{B}} a \vee^{\mathfrak{B}} b = \top^{\mathfrak{B}} \Leftrightarrow a \leqslant^{\mathfrak{B}} b$, for every Boolean algebra $\mathfrak{B}$ and all $a, b \in B$. It is algebraizable with equivalent purely-algebraic semantics BDMA (cf. [25] and Remark 3.60). In this way, the work [29] had provided us with the lattice of its extensions. We highlight just one of them here. This corresponds to NITSFBDMA and is relatively axiomatized by the single rule (3.33). It is constituted exactly by all rules admissible in the initial sequent calculus.

Conclusions

Thus, we have completely justified the key thesis of the book "equivalential versus algebraizable universal Horn theories" by arguing that any equivalential UHT does have an equivalent algebraic semantics of full value that is a prevariety of algebraic systems determined by the UHT uniquely up to rational equivalence (in our extended sense going back to [34]), while any algebraizable UHT just has an equivalent *purely-* algebraic semantics that is a prevariety of pure algebras determined by the UHT uniquely up to rational equivalence as well (but in the restricted original sense of [15]). In this way, we have done two parallel extensions implementing Mal'cev's paradigm "algebraic systems versus pure algebras". The first (preliminary) one consists in extending Mal'cev's conception of *rational equivalence* [15] from pure algebras to algebraic systems. The categorical characterization of our extended concept, going back to [34] and naturally generalizing Mal'cev's one [15], well justifies our primary extension. Then, our second (principal) extension, becoming essentially a secondary one, consists in extending the conception of equivalent algebraic counterpart of an UHT from pure algebras to algebraic systems as well. As we have argued, this extension respects the issue of "uniqueness of equivalent algebraic counterpart modulo rational equivalence". And what is more, once equivalence between UHTs preserves their metalogical features like extensions, relative axiomatizations, etc., such preservation can be applied to equivalential UHTs and their equivalent algebraic semantics like it has been applied to algebraizable UHTs and their equivalent purely-algebraic semantics in [24], though within the context of finitary universal Horn Logic alone. Since this finitary context is too restrictive, even if primary UHTs under consideration are finitary, as it has been argued in [34], we explore the issue involved within the general context of infinitary universal Horn Logic but with finitary signature symbols [34] (as opposed to [32], where infinitary symbols are admitted).[7]

[7]On the other hand, the latter limitation is crucial for the issue of *indiscernability* underlying the study [34], as it has been shown in [32].

Our justification of suggested extensions would not be complete without exemplifying our general elaboration by exploring several particular non-algebraizable equivalential UHTs naturally arizing in the literature on General Logic. We have chosen certain sequent calculi with structural rules which have been proved to be non-algebraizable in [24] and [25]. All these calculi are covered by the generic approach developed in [31]. On the other hand, as it has been proved in [32], all calculi of such a kind are finitely-equivalential. For these reasons, the selected four examples[8] seem to be most representative ones. Nevertheless, they do not at all exhaust all possible applications of our general study. And we encourage others to join us for providing more examples.

In general, both the quasivariety of simple quasi-ordered Stone algebras and the one of filtered Boolean De Morgan algebras have proved quite illustrative examples demonstrating that advanced methods of Universal Algebra are applicable to not merely algebraizable but equally equivalential logics. For studying the latter, it just suffices to follow Mal'cev's generic paradigm "algebraic systems versus pure abstract algebras" (cf. [16]).

Thus, once the concept of equivalent algebraic semantics results from that of equivalent purely-algebraic semantics by means of involving algebraic systems in place of pure algebras, we can eventually conclude that the notion of equivalential UHT results from the one of algebraizable UHT by extending pure algebras (as algebraic models) to algebraic systems.

In this connection, it is worth to highlight once more (cf. [34]) that the restricted "many-dimensional" formalism, adopted, e.g., in [26] and actually belonging to the so-called *abstract* (to be called more correctly *orthodox*) *algebraic logic* disables, in principle, both to cover prevarieties of algebraic systems (as opposed to those of pure algebras) and to treat equivalential UHTs within the context of any general theory of equivalence between UHTs like this has been implemented in [34]. On the other hand, it is such treatement that has enabled us to treat prevarieties of algebraic systems as equivalent algebraic semantics for equivalential UHTs of full value. This displays one of main crucial advantages of Mal'cev-style General Algebraic Logic developed in [24], [25], [32], [34], [35] and here over the orthodox algebraic logic[9], the main activity of which could not but be reduced by now, under these circumstances, to unmotivated associations of pure algebras with logical systems without taking the least care of any reasonable justification of such essentially blank algebraizations. It

[8] Recall that Section 3.2 covers two examples at once.

[9] Properly speaking, orthodox paradigms within Algebraic Logic arise from the orthodox prejudice within the classical modern Universal Algebra [5], [6], [11], as opposed to [16], declining anything not reducible to pure algebras.

is then not surprising that certain quite dishonest pseudo-researches in orthodox algebraic logic have been inclined to plagiarize somebody else's work, in particular, the main result of [35], about which (including the proof) they learned from the author in 1995.

And what is more, the current study demonstrates the crucial value of the general first-order formalism, adopted initially in [24] just to cover sequent calculi of any kind (see also [33] in this connection). This demonstrates how unexpected are sometimes ways of developement of fundamental science, when some tool proposed for solving one task becomes quite useful for solving another problem. And what is more, this expands both scopes and paradigms of Algebraic Logic in general, once algebraic systems are much wider than pure algebras.

On the other hand, equivalence between UHTs suggested in [34] preserves not solely extensions but equally many other metalogical points, in particular, Deduction Theorem (cf. Section A.1 in this connection). In this way, potential applications of the expansion developed here go far beyond the context of the present monograph. And we strongly encourage others to join us for further investigating all such issues.

Among other things, it is noteworthy that the quite non-trivial algebraic problem of finding all implicational classes of quasi-ordered De Morgan lattices remains still open.

Bibliography

[1] A. R. Anderson and N. D. Belnap. *Entailment, vol. 1*. Princeton University Press, Princeton, 1975.

[2] R. Balbes and P. Dwinger. *Distributive Lattices*. University of Missouri Press, Columbia (Missouri), 1974.

[3] N. D. Belnap, Jr. A useful four-valued logic. In J. M. Dunn and G. Epstein, editors, *Modern uses of multiple-valued logic*, pages 8–37. D. Reidel Publishing Company, Dordrecht, 1977.

[4] S.L. Bloom. Some theorems on structural consequence relations. *Studia Logica*, 34:1–9, 1975.

[5] S. Burris and H. P. Sankappanavar. *A Course in Universal Algebra*. Springer-Verlag, New York, 1981.

[6] P. M. Cohn. *Universal Algebra*. D. Reidel Publishing Company, Dordrecht, 2nd edition, 1981.

[7] M. Dummett. A propositional calculus with denumerable matrix. *Journal of Symbolic Logic*, 24:97–106, 1959.

[8] G. Gentzen. Untersuchungen über das logische Schliessen. *Mathematische Zeitschrift*, 39:176–210, 405–431, 1934.

[9] G. Grätzer. *Lattice Theory. First concepts and distributive lattices*. W. H. Freeman and Company, San Francisco, 1971.

[10] G. Grätzer. *General Lattice Theory*. Akademie-Verlag, Berlin, 1978.

[11] G. Grätzer. *Universal Algebra*. Springer-Verlag, Berlin, 2nd edition, 1979.

[12] J. A. Kalman. Lattices with involution. *Transactions of the American Mathematical Society*, 87:485–491, 1958.

[13] S. C. Kleene. *Introduction to metamathematics.* D. Van Nostrand Company, New York, 1952.

[14] J. Loś and R. Suszko. Remarks on sentential logics. *Indagationes Mathematicae*, 20:177–183, 1958.

[15] A. I. Mal'cev. The structural characterization of certain classes of algebras. *Doklady Akademii Nauk SSSR*, 120:29–32, 1958. In Russian.

[16] A. I. Mal'cev. *Algebraic systems.* Springer Verlag, New York, 1965.

[17] G. C. Moisil. Recherches sur l'algèbre de la logique. *Annales Scientifiques de l'Université de Jassy*, 22:1–117, 1935.

[18] G. Priest. The logic of paradox. *Journal of Philosophical Logic*, 8:219–241, 1979.

[19] T. Prucnal and A. Wroński. An algebraic characterization of the notion of structural completeness. *Bulletin of the Section of Logic*, 3:30–33, 1974.

[20] A. P. Pynko. A structural semantic approach to constructing propositional logical systems. Preprint 1815, Space Research Institute of Russian Academy of Sciences, Moscow, February 1992. In Russian.

[21] A. P. Pynko. Algebraic study of Sette's maximal paraconsistent logic. *Studia Logica*, 54(1):89–128, 1995.

[22] A. P. Pynko. Characterizing Belnap's logic via De Morgan's laws. *Mathematical Logic Quarterly*, 41(4):442–454, 1995.

[23] A. P. Pynko. On Priest's logic of paradox. *Journal of Applied Non-Classical Logics*, 5(2):219–225, 1995.

[24] A. P. Pynko. Definitional equivalence and algebraizability of generalized logical systems. *Annals of Pure and Applied Logic*, 98:1–68, 1999.

[25] A. P. Pynko. Functional completeness and axiomatizability within Belnap's four-valued logic and its expansions. *Journal of Applied Non-Classical Logics*, 9(1/2):61–105, 1999. Special Issue on Multi-Valued Logics.

[26] A. P. Pynko. Implication systems for many-dimensional logics. *Reports on Mathematical Logic*, 33:11–27, 1999.

[27] A. P. Pynko. Implicational classes of De Morgan lattices. *Discrete mathematics*, 205:171–181, 1999.

[28] A. P. Pynko. Subprevarieties versus extensions. Application to the logic of paradox. *Journal of Symbolic Logic*, 65(2):756–766, 2000.

[29] A. P. Pynko. Implicational classes of De Morgan Boolean algebras. *Discrete mathematics*, 232:59–66, 2001.

[30] A. P. Pynko. A cut-free Gentzen calculus with subformula property for first-degree entailments in $\mathbf{LC}^*$. *Bulletin of the Section of Logic*, 32(3): 137–146, 2003.

[31] A. P. Pynko. Sequential calculi for many-valued logics with equality determinant. *Bulletin of the Section of Logic*, 33(1):23–32, 2004.

[32] A. P. Pynko. A relative interpolation theorem for infinitary universal Horn logic and its applications. *Archive for Mathematical Logic*, 45:267–305, 2006.

[33] A. P. Pynko. Many-place sequent calculi for finitely-valued logics. *Logica Universalis*, 4(1):41–66, 2010.

[34] A. P. Pynko. *Equivalent universal Horn theories. General Algebraic Logic*. GRIN Verlag, Munich, 2018. ISBN 978-366-8-86946-2. URL `https://www.grin.com/document/452732`.

[35] A. P. Pynko. *Abstract sequent axiomatizations of finitary universal Horn theories. Abstract Proof Theory versus General Algebraic Logic*. Kindle Direct Publishing, 2019. ISBN 978-179-4-59554-5.

[36] A. M. Sette. On the propositional calculus P^1. *Mathematica Japonica*, 18:173–180, 1973.

Appendix A

Advances of equivalence between universal Horn theories

Here, we entirely follow Chapter 2 of [34].

First of all, remark that the following statement was implicitly proved in proving Corollary 2.8 of [34]:

Lemma A.1. *Let $\mathbb{T}_2$ and τ be as in Definition 2.1 of [34]. Suppose the conditions 2.1(v) as well as 2.7(v) of [34] hold. Then, the condition 2.8(iii) of [34] holds.*

This enables us to prove the following key observation of this appendix:

Proposition A.2. *Let $\mathbb{T}_2$ and τ be as in Definition 2.1 of [34]. Suppose the conditions 2.1(v) as well as 2.7(v) of [34] hold. Then, $\tau^{-1}[\vdash_2]$ is a structural L_1-consequence.*

Proof. Clearly, $\tau^{-1}[\vdash_2]$ is an L_1-consequence. Moreover, in view of the condition 2.8(iii) of [34] (cf. Lemma A.1), the structurality of $\tau^{-1}[\vdash_2]$ follows from the structurality of $\vdash_2$. This completes the argument. ■

As a consequence, we get the following auxiliary statement properly extending Lemma 2.5 of [34] to arbitrary translations:

Corollary A.3. *Let $\mathbb{T}_1$, $\mathbb{T}_2$ and τ be as in Definition 2.1 of [34]. Suppose the conditions 2.1(v) as well as 2.7(v) of [34] hold. Then,*

$$(\mathbb{T}_1 \subseteq \tau^{-1}[\vdash_2]) \Rightarrow (\vdash_1 \subseteq \tau^{-1}[\vdash_2]).$$

Proof. By Proposition A.2 and the primary definition of $\vdash_1$ as the least structural L_1-consequence including $\mathbb{T}_1$ (cf. Section 1.3 of [34]). ∎

Now we are in a position to argue the following appropriate extension of Proposition 2.4 of [34] to arbitrary (not necessarily purely relational) translations:

Proposition A.4. *Let $\mathbb{T}_1$, $\mathbb{T}_2$, τ and ρ be as in Definition 2.1 of [34]. Then, $\mathbb{T}_1$ and $\mathbb{T}_2$ are equivalent with respect to τ and ρ iff the conditions 2.1(iii-vi) as well as 2.7(v,vi) together with 2.4(i,ii) of [34] hold.*

Proof. The "only if" part immediately follows from Theorem 2.7 of [34]. To prove the converse, assume the conditions 2.1(iii-vi) as well as 2.7(v,vi) together with 2.4(i,ii) of [34] hold. Then, by 2.4(i), we have $\mathbb{T}_1 \subseteq \tau^{-1}[\vdash_2]$, so, by Corollary A.3, we get $\vdash_1 \subseteq \tau^{-1}[\vdash_2]$. This yields the metaimplication from left to right in 2.1(i). By symmetry, we equally obtain the metaimplication from left to right in 2.1(ii). These, by 2.1(iv,iii), imply the metaimplications from right to left in 2.1(ii,i), respectively. Thus, 2.1(i-vi) do hold, as required. ∎

Now, we are in a position to prove the following appropriate extension of Corollary 2.6 of [34] to arbitrary translations:

Corollary A.5. *Let $\mathbb{T}_1$, $\mathbb{T}_2$, τ and ρ be as in Definition 2.1 of [34]. Assume $\mathbb{T}_1$ and $\mathbb{T}_2$ are equivalent with respect to τ and ρ. Let $\mathbb{T}'$ be the universal Horn L_2-theory constituted by:*

(i) $\tau[\mathbb{T}_1]$;

(ii) $\{\tau[\rho(\Psi)] \to \Psi\} \cup \{\Psi \to \Phi | \Phi \in \tau[\rho(\Psi)]\}$, *for every* $\Psi \in \mathrm{At}_2(V^\infty)$;

(iii) $\{\tau(r) \to \Phi | \Phi \in \tau(r)[x_0/x_1] \setminus \tau(r)\}$, *for each constant* $r \in R_1$;

(iv) *for every* $\Psi \in \mathrm{At}_2(V^\omega)$ *and each constant* $c \in F_1$, *the following two rules:*

$$\begin{aligned} \{\Psi[x_1/\tau(c)]\} &\to \Psi[x_1/\tau(c)[x_0/x_1]], \\ \{\Psi[x_1/\tau(c)[x_0/x_1]]\} &\to \Psi[x_1/\tau(c)]. \end{aligned}$$

Then, $\vdash_2 = \vdash_{\mathbb{T}'}$.

In particular, for any $\omega \leqslant \lambda \leqslant \infty$, $\vdash_2$ is λ-universally axiomatizable iff $\vdash_1$ is.

Proof. We are going to prove that $\mathbb{T}_1$ and $\mathbb{T}'$ are equivalent with respect to τ and ρ with using Proposition A.4. First, 2.4(i) is by rules (i). Next, as 2.1(iii) holds, while $\mathbb{T}_1 \subseteq \vdash_1$, we have $\rho[\tau[\mathbb{T}_1]] \subseteq \vdash_1$. Moreover, by 2.1(ii,iv,v) and Theorem 2.7(v) of [34], we have $\rho[\tau[\rho(\Psi)]] \dashv\vdash_1 \rho(\Psi)$, for every $\Psi \in \mathrm{At}_2(V^\infty)$, while $\rho[\tau(r)] \vdash_1 \rho[\tau(r)[x_0/x_1]]$, for each constant $r \in R_1$, whereas $\rho(\Psi[x_1/\tau(c)]) \dashv\vdash_1 \rho(\Psi[x_1/\tau(c)[x_0/x_1]])$, for every $\Psi \in \mathrm{At}_2(V^\omega)$ and each constant $c \in F_1$. In this way, 2.4(ii) holds. Next, 2.1(iv) holds by rules (ii). Further, 2.1(v) holds by rules (iii). Finally, 2.7(v) holds by rules (iv). After all, 2.1(iii,vi) hold as such, while 2.7(vi) holds by Theorem 2.7(vi) of [34]. Thus, by Proposition A.4, $\mathbb{T}_1$ and $\mathbb{T}'$ are equivalent with respect to τ and ρ. Hence, by 2.1(ii), $\vdash_2 = \vdash_{\mathbb{T}'}$, as required. ■

A.1 Deduction Theorem for equivalent finitary UHTs

Here, we also follow Appendix B of [35].

The following result properly extends Theorem B.2 of [35] as well as its argumentation to arbitrary translations:

Theorem A.6. *Let $\mathbb{T}_1$, $\mathbb{T}_2$, τ and ρ be as in Definition 2.1 of [34]. Assume $\mathbb{T}_1$ and $\mathbb{T}_2$ are equivalent with respect to τ and ρ. Suppose $\mathbb{T}_2$ is finitary, while $\mathbb{T}_1$ has DT. Then, $\mathbb{T}_2$ has DT as well.*

Proof. Assume $\mathbb{T}_1$ has DT with respect to λ. In view of Proposition 2.3(i) of [34], ρ can be chosen to be finitary. Define a binary L_2-system λ' as follows. Consider any $r, r' \in R_2$. Take any $\overline{\Omega} \in \rho(r)^n$, where $n \triangleq |\rho(r)| \in \omega$, such that $\mathrm{img}\,\overline{\Omega} = \rho(r)$. Put $\lambda'(r, r') \triangleq \tau[\lambda(\overline{\Omega}, \rho(r')[x_j/x_{j+\mu_{L_2}(r)}]_{j\in\mu_{L_2}(r')})]$. Let $\overline{\varphi} \in \mathrm{Tm}_{F_2}(V^\infty)^{\mu_{L_2}(r)}$ and $\overline{\psi} \in \mathrm{Tm}_{F_2}(V^\infty)^{\mu_{L_2}(r')}$. Set

$$\sigma \triangleq [x_i/\varphi_i, x_{j+\mu_{L_2}(r)}/\psi_j]_{i\in\mu_{L_2}(r), j\in\mu_{L_2}(r')}.$$

Then, we have, for all $\Gamma \in \wp(\mathrm{At}_{L_2}(V^\infty))$:

$\Gamma \cup \{r(\overline{\varphi})\} \vdash_2 r'(\overline{\psi})$
$\rho[\Gamma] \cup \mathrm{img}(\rho\sigma)\overline{\Omega} \vdash_1 (\rho\sigma)[\rho(r')[x_j/x_{j+\mu_{L_2}(r)}]_{j\in\mu_{L_2}(r')}]$ by 2.1(ii) of [34]
$\rho[\Gamma] \vdash_1 \lambda((\rho\sigma)\overline{\Omega}, (\rho\sigma)[\rho(r')[x_j/x_{j+\mu_{L_2}(r)}]_{j\in\mu_{L_2}(r')}])$ by (B.7) of [35]
$\rho[\Gamma] \vdash_1 (\rho\sigma)[\lambda(\overline{\Omega}, \rho(r')[x_j/x_{j+\mu_{L_2}(r)}]_{j\in\mu_{L_2}(r')})]$ by (B.8) of [35]
$\tau[\rho[\Gamma]] \vdash_2 \tau[(\rho\sigma)[\lambda(\overline{\Omega}, \rho(r')[x_j/x_{j+\mu_{L_2}(r)}]_{j\in\mu_{L_2}(r')})]]$ by 2.1(i) of [34]
$\Gamma \vdash_2 \tau[(\rho\sigma)[\lambda(\overline{\Omega}, \rho(r')[x_j/x_{j+\mu_{L_2}(r)}]_{j\in\mu_{L_2}(r')})]]$ by 2.1(iv) of [34]
$\Gamma \vdash_2 (\tau\rho\sigma)[\tau[\lambda(\overline{\Omega}, \rho(r')[x_j/x_{j+\mu_{L_2}(r)}]_{j\in\mu_{L_2}(r')})]]$ by 2.8(iii) of [34]
$\Gamma \vdash_2 (\tau\rho\sigma)[\lambda'(r, r')]$ by definition of $\lambda'(r, r')$
$\Gamma \vdash_2 \sigma[\lambda'(r, r')]$ by 2.7(iv) and 1.10 with $V = V^{\mu_{L_2}(r)+\mu_{L_2}(r')}$
and $V' = \mathrm{Var}(\{r(\overline{\varphi})\} \to r'(\overline{\psi}))$ of [34]
$\Gamma \vdash_2 \lambda'(r(\overline{\varphi}), r'(\overline{\psi}))$ by definition of the mapping λ'

Thus, $\mathbb{T}_2$ has DT with respect to λ', as required. ■

Remark that the proof of Theorem A.6 is entirely constructive, while the construction of appropriate binary systems involves solely relational components of translations under consideration, like such was the case, when dealing with purely relational translations alone (cf. the constructive proof of Theorem B.2 of [35].)

Appendix B

Equivalent relational fragments of universal Horn theories

Here, we again follow the conventions adopted in Chapter 2 of [34].

Suppose $F_1 = F_2$. Let $\mathbb{T}$ be a universal Horn L_1-theory and τ a purely-relational (L_1, L_2)-translation. Put $\tau * \mathbb{T} \triangleq \tau[\mathbb{T}] \cup \{\tau(r) \to \Psi | r \in R_1, \mu_1(r) = 0, \Psi \in \tau(r)[x_0/x_1] \setminus \tau(r)\}$. (In case τ is parameter-less, we have $\tau * \mathbb{T} = \tau[\mathbb{T}]$.)

We start from proving the following general auxiliary statement useful in its own right (see, e.g., Appendix C).

Lemma B.1. *Suppose $F_1 = F_2$. Let $\mathbb{T}$ be a universal Horn L_1-theory, τ a purely-relational (L_1, L_2)-translation and ρ a purely-relational (L_2, L_1)-translation such that $\tau \circ \rho = \iota_2$, while $\rho[\tau(\Phi)] \dashv\vdash_{\mathbb{T}} \Phi$, for every $\Phi \in \mathrm{At}_1$, and $\rho(p) \vdash_{\mathbb{T}} \rho(p)[x_0/x_1]$, for every nullary $p \in R_2$ (in particular, ρ is parameter-less). Then, $\mathbb{T}$ and $\tau * \mathbb{T}$ are equivalent with respect to τ and ρ.*

Proof. With using Proposition 2.4 of [34]. First of all, the conditions 2.4(i) and 2.1(v) are trivial, by definition of $\tau * \mathbb{T}$. Moreover, as $\tau \circ \rho = \iota_2$, the condition 2.4(iv) holds too. Next, as $\rho(p) \vdash_{\mathbb{T}} \rho(p)[x_0/x_1]$, for every nullary $p \in R_2$, the condition 2.1(vi) holds as well. Finally, the fact that $\rho[\tau(\Phi)] \dashv\vdash_{\mathbb{T}} \Phi$, for every $\Phi \in \mathrm{At}_1$, implies, first, the condition 2.4(iii) together with the inclusion $\rho[\tau[\mathbb{T}]] \subseteq \vdash_{\mathbb{T}}$. Second, by Lemma 2.5(2.3) of [34] (with ρ instead of τ), we get, for each constant $r \in R_1$:

$$\rho[\tau(r)] \vdash_{\mathbb{T}} r = r[x_0/x_1] \vdash_{\mathbb{T}} \rho[\tau(r)][x_0/x_1] \vdash_{\mathbb{T}} \rho[\tau(r)[x_0/x_1]].$$

Thus, the condition 2.4(ii) equally holds, as required. ■

Applying Lemma B.1 with parameter-less $\rho = \iota_2$, we immediately get:

Theorem B.2. *Suppose $F_1 = F_2$, while $R_2 \subseteq R_1$. Let $\mathbb{T}$ be a universal Horn L_1-theory and τ an (L_1, L_2)-translation such that $\tau \upharpoonright L_2 = \iota_2$ (in which case τ is purely relational), while $\tau(\Phi) \dashv\vdash_{\mathbb{T}} \Phi$, for every $\Phi \in \mathrm{At}_1$. Then, $\tau * \mathbb{T}$ and $\mathbb{T}$ are equivalent with respect to ι_2 and τ.*

B.1 Application to disjunctive algebraic systems

During the rest of the monograph, like Chapter 2, we entirely follow conventions adopted in Section 1.1 and Appendix C of [35] except that α (like β) is supposed to be an arbitrary element of $\{\omega, \omega \setminus 1, 2, 2 \setminus 1\}$, in which case all the results, obtained therein for the case $\alpha \in \{\omega, \omega \setminus 1\}$, are extended to the case $\alpha \in \{2, 2 \setminus 1\}$ with exactly the same argumentation.

During the rest of this appendix, we suppose that $\beta \subseteq 2$. We start from presenting the following refined version of Lemma C.3 of [35] proved therein for parameter-less binary L-systems λ alone, in which case corresponding translations χ_λ are parameter-less as well. Namely, by Theorem B.2, we have:

Lemma B.3. *Let $\mathbb{S}$ be an L-sequent calculus of type $[\alpha, \beta \cup (\omega \setminus 2)]$ such that $\chi_\lambda(\Phi) \dashv\vdash_{\mathbb{S}} \Phi$, for every $\Phi \in \mathrm{Seq}_L^{[\alpha,\beta\cup(\omega\setminus 2)]}$. Then, $\mathbb{S}$ and $\chi_\lambda * \mathbb{S}$ are equivalent with respect to χ_λ and $\iota_{L^\alpha_\beta}$.*

Combining (C.3), Lemma C.2 of [35] with Lemma B.3, we equally get the following refined version of Theorem C.4 proved therein for parameter-less λ as well:

Theorem B.4. *Let K be a class of λ-disjunctive L-structures and $\mathbb{S}$ an L-sequent calculus of type $[\alpha, \beta \cup (\omega \setminus 2)]$ such that $\vdash_{\mathbb{S}} = \vdash_{\mathsf{K}[\alpha,\beta\cup(\omega\setminus 2)]}$. Then, $\vdash_{\chi_\lambda * \mathbb{S}} = \vdash_{\mathsf{K}[\alpha,\beta]}$.*

Appendix C

Sequent calculi for complementary algebraic systems

Given any L-structure $\mathcal{A}$, we have the *complementary* L-structure $\overline{\mathcal{A}}$ with the same underlying algebra $\mathfrak{A}$ and relations defined as follows: for each $r \in R$, put $r^{\overline{\mathcal{A}}} \triangleq A^{\mu_L(r)} \setminus r^{\mathcal{A}}$. Clearly:

$$\overline{\overline{\mathcal{A}}} = \mathcal{A}. \tag{C.1}$$

Given a class of L-structures K, put $\overline{\mathsf{K}} \triangleq \{\overline{\mathcal{A}} | \mathcal{A} \in \mathsf{K}\}$.

Consider the purely relational parameter-less *side-inversion* $(L^{\alpha}_{\beta}, L^{\beta}_{\alpha})$-translation $\varsigma^{\alpha}_{\beta}$ defined by (where $m \in \alpha$, $n \in \beta$, $\overline{p} \in R^m$, $\overline{r} \in R^n$):

$$\varsigma^{\alpha}_{\beta}(\rightarrowtail^{\overline{p}}_{\overline{r}}) \triangleq \{(r_j(\overline{x}(\mu_L(\overline{p}) + \mu_L(\overline{r} \restriction j), \mu_L(r_j))_{j \in n} \\ \rightarrowtail (p_i(\overline{x}(\mu_L(\overline{p} \restriction i), \mu_L(p_i))_{i \in m}\}.$$

Notice that:

$$\varsigma^{\beta}_{\alpha} \circ \varsigma^{\alpha}_{\beta} = \iota_{L^{\alpha}_{\beta}}. \tag{C.2}$$

Moreover, given any L-structure $\mathcal{A}$, we have:

$$(\varsigma^{\beta}_{\alpha})^{-1}[\mathcal{A}^{[\alpha,\beta]}] = \overline{\mathcal{A}}^{[\beta,\alpha]}. \tag{C.3}$$

Then, by Theorem 2.10(v)⇒(i) of [34], (C.1) and (C.3) immediately yield:

Proposition C.1. *Let* K *be a class of* L*-structures. Then,* $\vdash_{\mathsf{K}^{[\alpha,\beta]}}$ *and* $\vdash_{\overline{\mathsf{K}}^{[\beta,\alpha]}}$ *are equivalent with respect to* $\varsigma^{\alpha}_{\beta}$ *and* $\varsigma^{\beta}_{\alpha}$*.*

Applying (C.2) and Lemma B.1 with parameter-less $\rho = \varsigma^{\beta}_{\alpha}$ and $\tau = \varsigma^{\alpha}_{\beta}$, we get:

Lemma C.2. *Let $\mathbb{S}$ be an L-sequent calculus of type $[\alpha, \beta]$. Then, $\mathbb{S}$ and $\varsigma^{\alpha}_{\beta}[\mathbb{S}]$ are equivalent with respect to $\varsigma^{\alpha}_{\beta}$ and $\varsigma^{\beta}_{\alpha}$.*

Proposition C.1 and Lemma C.2 directly imply:

Theorem C.3. *Let K be a class of L-structures and $\mathbb{S}$ an L-sequent calculus of type $[\alpha, \beta]$. Assume $\vdash_{\mathsf{K}[\alpha,\beta]} = \vdash_{\mathbb{S}}$. Then, $\vdash_{\overline{\mathsf{K}}[\beta,\alpha]} = \vdash_{\varsigma^{\alpha}_{\beta}[\mathbb{S}]}$.*

C.1 Application to conjunctive algebraic systems

Let λ be a binary L-system. An L-structure $\mathcal{A}$ is said to be λ-*conjunctive*, whenever $\overline{\mathcal{A}}$ is λ-disjunctive (cf. Appendix C of [35]). From now on, $\alpha \in \{2, 2 \setminus 1\}$. Put $\chi^{\lambda} \triangleq \varsigma^{\beta}_{\alpha} \circ \chi_{\lambda} \circ \varsigma^{\alpha \cup (\omega \setminus 2)}_{\beta}$.

Combining Proposition 2.9 of [34] with Lemma C.2 of [35], Theorem B.4, (C.1), Proposition C.1 and Theorem C.3, we immediately obtain the following "dual version" of both Lemma C.2 of [35] and Theorem B.4:

Theorem C.4. *Let K be a class of λ-conjunctive L-structures and $\mathbb{S}$ an L-sequent calculus of type $[\alpha \cup (\omega \setminus 2), \beta]$ such that $\vdash_{\mathsf{K}[\alpha \cup (\omega \setminus 2),\beta]} = \vdash_{\mathbb{S}}$. Then, $\vdash_{\mathsf{K}[\alpha,\beta]} = \vdash_{\chi^{\lambda} * \mathbb{S}}$ is equivalent to $\vdash_{\mathbb{S}}$ with respect to $\iota_{L^{\alpha}_{\beta}}$ and χ^{λ}.*

In case λ is parameter-less, χ^{λ} is parameter-less as well, in which case $\chi^{\lambda} * \mathbb{S} = \chi^{\lambda}[\mathbb{S}]$.

www.ingramcontent.com/pod-product-compliance
Lightning Source LLC
Chambersburg PA
CBHW071232220526
45468CB00002B/816